Praise for *Blockchain and Web3*

T0311378

"A thoughtful guide to the role blockchain and crypto assets play in the world-changing internet transformation – and how one accelerates the other."

— Anthony Scaramucci, Founder & Managing Partner of SkyBridge

"Provides a colorful account of how things like gaming, blockchain, NFTs, AR/VR, DAOs, and DeFi have converged and ultimately presented to us this whole package called metaverse."

— Clay Lin, Chief Information Security Officer (CISO), World Bank Group

"Ma and Huang provide the essential handbook on the transformative power of Web 3 – taking you on a ride from the basic fundamentals of the blockchain protocols to the vast possibilities of the metaverse – and the immense impact it could bring. An educational and entertaining must-read for anyone interested in the next, programmable and immersive, web."

— Lila Tretikov, deputy Chief Technology Officer (CTO), Microsoft

"An essential breakdown of the most important recent developments in the blockchain space. Unlike many other writers, Ma and Huang look beyond mere financial speculation to uncover the true productive potential of blockchains, smart contracts, DAOs, DeFi, the metaverse, and more."

— Neel Mehta, Author of *"Bubble or Revelation?: The Future of Bitcoin, Blockchains, and Cryptocurrencies"*

"A remarkable convergence of digital economy with blockchain and Web3, depicting the true merits of the metaverse in relinquishing the impending daunted horizons of the information technology."
— **Mehdi Paryavi DEA®, Chairman, the International Data Center Authority (IDCA)**

"A clear picture of the complex ecosystem that enables the next-generation internet. Readers will become "Web3 smart" netizens, educated participants, and even adept game changers."
— **William Zhang, Security Architecture Lead, World Bank Group**

"Provides a valuable window into metaverse and covers the important building blocks for a trusted metaverse."
— **Yale Li, Chairman, Cloud Security Alliance – Greater China Region (CSA GCR)**

"Metaverse, Web3, and blockchain are among the cutting-edge technologies of the new digital economy. Focused on security, privacy, and data governance, this book discusses the paramount aspects of how these new technologies are used in the real world."
— **Yao Qian, Ex-Head of China's Digital Yuan Effort, now Director of the Science and Technology Supervision Bureau of China Securities Regulatory Commission**

BLOCKCHAIN AND
WEB3

WINSTON MA
KEN HUANG

BLOCKCHAIN AND WEB3

BUILDING THE CRYPTOCURRENCY, PRIVACY, AND SECURITY FOUNDATIONS OF THE METAVERSE

WILEY

To Angela – I love you dearly.

– Winston Ma

To Queenie Ma, Grace Huang, and Jerry Huang, for your unwavering love, support, and encouragement.

– Ken Huang

Contents

Foreword

Great Leap Forward into Web3

In September 2017, JPMorgan Chase (America's largest investment bank firm) CEO Jamie Dimon called Bitcoin a "fraud." "It's worse than tulip bulbs. It won't end well. Someone is going to get killed," Dimon said. Now, its wealthiest clients can invest in the asset on the bank's own platform. The dramatic shift of JPMorgan is a significant milestone for the Bitcoin, as well as broad cryptocurrencies, as an asset class.

Various financial institutions like JPMorgan, both on Wall Street and in international governments, have had a very complicated relationship with cryptocurrency as a whole since Bitcoin (together with blockchain technology) first crashed onto the world stage 10 years ago. But as digital finance has accelerated by the pandemic, institutions actively explore new avenues to get involved in the crypto space. This institutional adoption has benefited not only Bitcoin but also the entire crypto asset industry, helping break barriers all across the board.

For example, Morgan Stanley, which has the country's largest wealth management unit with nearly $5 trillion in assets under management and advisement, has created Bitcoin products on its platform for ultra-high-net-worth investors. U.S. Bank, which is part of U.S. Bancorp, the fifth-largest bank in America announced a new cryptocurrency custody product. Goldman Sachs and other Wall Street banks have started looking into how to use bitcoin as collateral for cash loans to institutions.

Going beyond crypto investments and trading, Bank of America recently released a major research report, stating that they see a massive opportunity in the *Metaverse*, and that it could spur the wider adoption of the crypto industry. One of their top strategists said that he expects large traditional financial companies to enter the space once crypto assets gain wider adoption and usage in the metaverse, and it will finally cause cryptocurrencies to start being used widely for transactions.

In short, Bitcoin, crypto assets, and decentralized technologies (including blockchain) are much more than its financial origins, and this is becoming apparent around the world. Instead, it's about a new, better internet known as Web3. Digital assets and Web3 projects are radically changing how we invest, strategize business models, and deploy products and services. These projects have not only disrupted the thinking of institutional and professional investors, but also have inspired global brands and entrepreneurs to develop new products and services for both the physical and virtual worlds.

Blockchain is the backbone of Web3, which may be the next major platform in computing after the World Wide Web (Web1.0) and mobile internet (Web2.0). It is poised to revolutionize every industry and function, from finance and health care to media entertainment and real estate, creating trillions in new value – and the radical reshaping of society.

Ma has produced a terrific and highly accessible field guide to understanding how the digital economy is accelerating in the Web3 metaverse. A nationally certified software programmer as early as 1994, Ma has published many books on global tech revolution, including *The Hunt for Unicorns: How Sovereign Funds Are Reshaping Investment in the Digital Economy* (2020) and *The Digital War – How China's Tech Power Shapes the Future of AI, Blockchain, and Cyberspace* (2021). For both, I made similar book recommendations to major financial institutions, asset managers, hedge funds, as well as other key players and stakeholders.

As an investor, attorney, author, and adjunct professor in the global digital economy, Ma addressed the crypto-based

Web3 metaverse from various perspectives, together with his co-author Ken Huang, a blockchain security expert. The authors' extensive, hands-on involvement in the deals and operations of this mystical world lends vibrancy as they recount practical, illustrative examples in a non-pedantic style. Together, their unique perspectives and differing approaches have produced a nuanced roadmap to the little-known past and exciting prospects of blockchain internet.

Sometimes a book sheds light on a little-known but powerful force. Sometimes it is timely because it catches the world at an inflection point. Rarely does a book accomplish both. With the arrival of *Blockchain and Web3* from Winston Ma and Ken Huang, we have that rare beast: a book that, against the backdrop of the world-altering coronavirus epidemic, provides a thoughtful guide to the role blockchain and crypto assets play in the world-changing internet transformation – and how one accelerates the other.

Anthony Scaramucci
Founder and Managing Partner of SkyBridge

The Opportunities and Challenges of Metaverse

Covid-19 has accelerated digital transformation across the globe, from virtual meetings and electronic signatures to digital payments and remote supervision, just to name a few. In the meantime, another strong force is shaping up the next-generation internet, or Web3. We often hear the ingredients of Web3: blockchain, decentralized finance or DeFi, nonfungible tokens (NFTs), and most recently the metaverse. We hear the opportunities as well as challenges these emerging technologies bring about to individuals, organizations, and regulators, and become anxious every day.

As a fast-evolving field, Web3 and its enabling technologies are developing very rapidly. This makes it hard for people to stay current and make informed decisions as to how to take advantage of the opportunities, how to manage the risks, or simply, how to participate.

Luckily, Winston Ma and Ken Huang have provided readers of this book a very detailed picture of the current Web3 landscape. Having been practitioners in this space for many years, Winston and Ken give us a vivid account of the major events and players in each of the fields in the Web3 ecosystem, from technology innovation, new business models, participation by established companies whose current business may be disrupted, the various types and stances of cybersecurity hacks, to reactions from government regulators. This holistic view is beneficial for people to understand the development of this dynamic and complex ecosystem before they can take informed actions.

The year 2021 was marked as the year of the NFTs, when it became a buzzword for the masses and brought us landmark deals worth multimillion dollars. But many people do not understand what exactly they are getting into when they purchase an NFT generated from things like digital art. The recent story of the avid NFT collector, who paid $2.9 million for an NFT in 2021 but was not able to even get a bid close to $10,000 a year later, shows that people have different perceptions of what an NFT represents and what its intrinsic value is. Chapter 5 of the book provides the audience with useful information on this topic.

As Facebook changed its name to Meta in 2021, and Microsoft acquired gaming company Activision Blizzard for $68.7 billion in early 2022, many are puzzled about the value proposition of the metaverse, and what it means for them. The book provides a colorful account of how things like gaming, blockchain, NFTs, AR/VR, DAOs, and DeFi have converged and ultimately presented to us this whole package called metaverse. The chapters also present the challenges and opportunities that metaverse faces, prompting the audience to think about what these mean for their organization and for themselves.

Congratulations to Winston and Ken on a comprehensive and easy-to-read book that offers so much information and presents so many intriguing open questions for the audience to ponder and act on. Their research will elevate the level of understanding of Web3 by the blockchain and fintech communities and trigger actions that will help shape the next-generation internet for the benefit of humanity!

Clay Lin
Chief Information Security Officer (CISO)
World Bank Group

Blockchain: The Building Blocks of a Trusted Metaverse

Metaverse was predicted 30 years ago in Neal Stephenson's novel *Snow Crash,* where people interact as avatars within a high-definition virtual environment projected onto special goggles. Today, new digital technologies like blockchain will gradually join up and form the building blocks of the future metaverse, which could be the next generation of internet capable of transmitting 3D holograms and a lot more.

However, there are numerous potential obstacles – from technological and economical to political, security, and many other aspects – we must overcome to pave the way of the metaverse. From cloud computing's perspective, the majority of metaverse platform components will have to run on a secured cloud environment, which enforces zero trust and embraces blockchain innovations such as privacy preserving computing, decentralized storage, and decentralized identity as described in this book.

This book provides a valuable window into metaverse and covers the important building blocks for a trusted metaverse. Particularly, it explains blockchain as a critical technology to converge with metaverse. Cryptocurrencies, DeFi, NFT, gaming tokens, and other usage scenarios are discussed extensively in the book. Security and privacy have always been challenges to the internet and the digital world, and fortunately, they are paid full attention in the book as well. I have no doubt that you would enjoy state-of-the-art knowledge and insights on the metaverse from the book, whether you are a businessperson, tech investor, technical professional, government official, or student at college.

The authors of this book are senior experts Ken Huang and Winston Ma in the industry and academia. For many years, I have been very impressed with Ken's research leadership as VP of Research at CSA GCR. Being a recognized technology leader in blockchain field, Ken has published many standards, white papers, and training contents. In 2021, he won the award of "60 Blockchain Leaders" in China. I truly believe that no one else could share the convergence of blockchain and metaverse better than Ken and Winston.

Happy Reading,
Yale Yuhang Li
Foreign Member, Ukrainian Academy
of Engineering Sciences
Chairman, Cloud Security Alliance –
Greater China Region (CSA GCR)
Seattle, Washington USA

Acknowledgments

Winston Ma

In the middle of 1990s, the early days of China's tech and internet boom, I majored in electronic materials and semiconductor physics at Fudan University in Shanghai. Aiming for graduate studies in the United States, I diligently studied English for the TOEFL and GRE exams, and I also took a national exam for a professional certificate that is no longer relevant two decades later – "software programmer."

Back then, China had so few software programmers that the central government organized national qualification exams to encourage the young generation to study computer science. Sensing the tremendous potential of China's tech revolution, I sat in a one-day exam to solve coding problems in C, Fortran, and Pascal languages before I became a "nationally certified software programmer." Today, however, those programming languages are "old" for coding, and there is no need for such a national exam because numerous college students graduate from computer science majors, driven by the mobile internet boom started last decade and the Web3 revolution emerging post-Covid-19 pandemic.

That's why my 2022 book focuses on *Blockchain and Web3*, after my pentalogy on China's digital revolution and tech power in the previous five years, because we are entering into a new global era of digital transformation. A book on such a complex and fast-moving topic would not have been possible if I had not been blessed to partner with an industry leader like Ken, who has over 20 years of cybersecurity and blockchain technology experience in cloud security, identity and access management, and PKI and date encryption.

****

My deepest thanks go to Dr. Rita and Gus Hauser, the New York University (NYU) School of Law, and John Sexton, the legendary dean of NYU Law School when I was pursuing my LL.M degree in Comparative Law. My PE/VC investing, investment banking, and practicing attorney experiences all started with the generous Hauser scholarship in 1997. During his decade-long tenure as the president of NYU, John kindly engaged me at his inaugural President's Global Council as he developed the world's first and only GNU (global network university). My NYU experience was the foundation for my future career as a global professional working in the cross-border business world.

My sincere appreciation to both Mr. Lou Jiwei and Dr. Gao Xi-qing, the inaugural chairman and president of China Investment Corporation (CIC), for recruiting me at its inception. One of the most gratifying aspects of being part of CIC is the opportunity to be exposed to a wide range of global financial markets' new developments. The unique platform has brought me to the movers and shakers everywhere in the world, including Silicon Valley projects that linked global tech innovation with the Chinese market.

The same thanks go to Chairman Ding Xue-dong and President Li Ke-ping, who I reported to at CIC in recent years. Similarly, thanks to Linda Simpson, senior partner at the New York headquarters of Davis Polk & Wardwell, and Santosh Nabar, managing director at the New York headquarters of JPMorgan. Those two former bosses on Wall Street gave me a foundation to develop a career in the global capital markets.

Many thanks to Mr. Jing Liqun, president of Asian Infrastructure Investment Bank (AIIB) and formerly the supervisory chairman of CIC. He educated me about the works of Shakespeare, as well as guiding me professionally. The readings of

Hamlet, Macbeth, and *King Lear* improved my English writing skills, and hopefully the writing style of this book is more interesting and engaging than my previous finance textbook *Investing in China.*

For such a dynamic book topic, I benefited from the best market intelligence from a distinctive group of institutional investors, tech entrepreneurs, and business leaders at the World Economic Forum (WEF), especially the fellows at the Council on Long-Term Investing, the Council for Digital Economy and Society, and the Young Global Leaders (YGL) community. Professor Klaus Schwab, founder and executive chairman of the World Economic Forum, has a tremendous vision of a sustainable, shared digital future for the world, which is an important theme of this book.

The WEF Council on Long-Term Investing has gathered the most forward-thinking leadership from major sovereign wealth funds and public pensions, and I learned so much from the dynamic discussions with them for this book's coverage on the sovereign digital currency. They include Alison Tarditi (CIO of CSC, Australia), Adrian Orr (CEO of NZ Super, New Zealand), Gert Dijkstra (chief strategy of APG, Netherlands), Hiromichi Mizuno (CIO of GPIF, Japan), Jagdeep Singh Bachher (CIO of UC Regents, USA), Jean-Paul Villain (director of ADIA, UAE), Lars Rohde (CEO of ATP, Denmark), Lim Chow Kiat (CEO of GIC, Singapore), Reuben Jeffery (CEO of Rockefeller & Co., USA), and Scott E. Kalb (CIO of KIC, Korea).

My gratitude goes to many other outstanding friends, colleagues, practitioners, and academics who provided expert opinions, feedback, insights, and suggestions for improvement. For anecdotes, pointers, and constant reality checks, I turned to them because they were at the front line of industry and business practices. I would particularly like to thank my partners at CloudTree Ventures (a VC fund focusing on the technologies driving interactive entertainment and the metaverse), Trevor Barron, Jeffery Schoonover, and Adam Smith, as well as the friends at Capgemini, where I am a member of its advisory board, including Cornelia Schaurecker (Global Group Director AI & Big Data of Vodafon), Lila Tretikov (deputy CTO of Microsoft), and Mishka Dehghan (SVP Strategy, Product, & Solutions Engineering of T-Mobile).

On its journey from a collection of ideas and themes to a coherent book, the manuscript went through multiple iterations and a meticulous editorial and review process by the John Wiley team led by the book commissioning editor Gemma Valler. Our long-term collaboration started with my 2016 book, *China's Mobile Economy*. During the pandemic, we released *The Hunt for Unicorns: How Sovereign Funds Are Reshaping Investment in the Digital Economy* (2020) and *The Digital War – How China's Tech Power Shapes the Future of AI, Blockchain and Cyberspace* (2021). The managing editor Purvi Patel and copyeditor Cheryl Ferguson contributed substantially to the final shape of the book. Special thanks to Gladys Ganaden for her design of the book cover and figures.

And last in the lineup but first in my heart, I thank my wife, Angela Ju-hsin Pan, who gave me love and support. You are a true partner in helping me frame and create this work. Thanks for your patience while I wrecked our weekends and evenings working on this book.

Ken Huang

At the end of 2016, I resigned from my role overseeing blockchain technology strategic research and fintech product development at Huawei. The main reason for my resignation is that decentralization technology and its associated innovation cannot happen inside Big Tech companies.

That is the main theme of this book. As I worked with my co-author, Winston Ma, to develop the contents of this book, it became even more clear that blockchain technology innovations will happen in a metaverse, led by many small startups.

I am very thankful for Winston Ma, who has come up with the initial idea of the book and developed the book contents as decentralization technology and tokenomics innovations have sped up during the global pandemic.

I am also very much indebted to Sally Gao, an alumna of McKinsey and Company, Chinese tech VC Sinovation Ventures, and Columbia University. She has contributed all figures of the book, created the Glossary section, and translated some of my contents from Chinese to English.

I certainly enjoyed long walks and deep discussions of Blockchain, DeFi, NFT, and Metaverse ideas with my daughter, Grace Huang, who is a product manager of PIMCO, an American investment management firm. This upcoming generation is promising, as lots of the good points from this book were the results of discussions with my daughter. Thank you to my wife, Queenie Ma, and my son, Jerry Huang (who is completing his master's degree and is also working as a teacher assistant for a blockchain course at Georgia Institute of Technology) for their love and encouragement as well as insightful discussions about the book.

I am also very grateful to the following individuals for forming my view of decentralization, blockchain security, privacy, DeFi, and metaverse applications in the past few years, including and in no particular order:

- Michael Casey, chief content officer of CoinDesk, for insightful discussion on privacy and self-sovereign identity
- Vitalik Buterin, co-founder of Ethereum, for discussion on blockchain scaling solution, privacy, sharding, layer 2, and many other topics
- Dr. Xiao Feng, chairman of Wanxiang Blockchain, for his support and comments on my previous book on blockchain security and continued discussion afterward
- Dr. Yao Qian, ex-head of China's Digital Yuan (CBDC) Effort and now director of the Science and Technology Supervision Bureau of the China Securities Regulatory Commission, for many discussions on the original CBDC design and related issues and concerns

- Professor Whitfield Diffie of Stanford University, for good discussions on cryptography and privacy
- Professor Jim Waldo of Harvard, who taught me at the Harvard Kennedy School executive program on Cybersecurity: The Intersection of Policy and Technology
- Clay Lin, CISO of World Bank Group, for many discussions on blockchain and his support for my previous book
- Yale Lee, chair of Cloud Security Alliance (CSA)-GCR, for collaborative works on blockchain security white papers published by CSA

About the Authors

Winston Wenyan Ma, CFA & Esq.

Winston Ma is an investor, attorney, author, and adjunct professor in the global digital economy. He is a co-founder and managing partner of CloudTree Ventures, a seed-to-early-growth-stage venture capital firm empowering interactive entertainment companies. He is currently the board chairman of Nasdaq-listed MCAA, a European tech SPAC, an advisory board member of Capgemini, and an adjunct professor at NYU Law School.

Most recently for 10 years, he was managing director and head of North America Office for China Investment Corporation (CIC), China's sovereign wealth fund. At CIC's inception in 2008, he was among the first group of overseas hires by CIC, where he was a founding member of both CIC's Private Equity Department and later the Special Investment Department for direct investing (head of CIC North America office 2014–2015). He had leadership roles in global investments involving financial services, technology (TMT), energy and mining sectors, including the setup of West Summit Capital, a cross-border growth capital fund in Silicon Valley – CIC's first overseas tech investment.

Prior to that, Ma served as the deputy head of equity capital markets at Barclays Capital, a vice president at JPMorgan investment banking, and a corporate lawyer at Davis Polk & Wardwell LLP.

A nationally certified Software Programmer as early as 1994, Ma is the author of the books *China's Mobile Economy* (2016), *Digital Economy 2.0* (2017 Chinese), *The Digital Silk Road* (2018

German), *China's AI Big Bang* (2019 Japanese), and *Investing in China* (2006). His new books are *The Hunt for Unicorns: How Sovereign Funds Are Reshaping Investment in the Digital Economy* (2020) and *The Digital War – How China's Tech Power Shapes the Future of AI, Blockchain, and Cyberspace* (2021).

Ma earned his MBA from the University of Michigan Ross Business School (Beta Gamma Sigma) and his Master of Law from the New York University School of Law (Hauser Global Scholar). He earned bachelor of science and bachelor of law degrees from Fudan University in Shanghai, China. He was selected a 2013 Young Global Leader at the World Economic Forum (WEF). In 2014 he received the NYU Distinguished Alumni Award.

Ken Huang

Ken Huang is CEO of *DistributedApps* and chair of Blockchain Security Working Group for *Cloud Security Alliance* in the Great China Region (CSA GCR). Over the past 18 years, he has worked on application security, identity access management, cloud security, and blockchain for the fintech industry and health care industry. He is a certified CISSP (Certified Information Systems Security Professional) since 2007 and authored a *Blockchain Security Technical Guide,* published by China Machine Press in 2018.

As CEO of *DistributedApps,* he provides cybersecurity consulting services on blockchain and AI for startup companies globally. Prepandemic, he has been nominated to judge blockchain and AI startup contests organized by Google, Softbank, and Stanford University in 2018. As part of W3C Credentials Community Group Member, he provided his comments for NIST 800-63 documents on identity management.

As chair of *CSA GCR,* he has worked with top security experts in the blockchain space to create more than 10 white papers or technical guides on blockchain security. As a devoted member of Cloud Security Alliance, he was a reviewer of *Blockchains in the Quantum Era* published February 2021. He was also the lead author of the Cloud Security Alliance document titled *Crypto-Asset Exchange Security Guidelines* published April 2021.

He is a blockchain advisor for Timeraiders.io, an NFT game project in the UK. He is also an advisor for KnownSec, a Hong Kong–based cybersecurity firm. In January 2021, he completed his Harvard Kennedy School Executive Program on Cybersecurity.

He has been invited to speak at numerous conferences globally covering blockchain, AI, and security, including Davos World Economic Forum, CoinDesk Consensus, IEEE, ACM, Worldbank, Stanford University, UC Berkeley, Bank of China, and Huawei. His speeches on layers of new technology and their convergence during Davos WEF 2020 were widely reported in many Chinese crypto media and used as original input in the first two chapters of this book.

****

The authors can be reached on Linkedin for comments and feedback on *Blockchain and Web3: Building the Cryptocurrency, Privacy, and Security Foundations of the Metaverse.*

Preface

For the vast majority of those watching the rapid rise of cryptocurrency (with blockchain), its emergence has been something of a curious novelty. The Russia–Ukraine war that started in February 2022 unexpectedly shines a spotlight on cryptocurrency, illustrating distract concepts like fast payment, decentralized network, and nonfungible token (NFT) in live, dramatic contexts.

In a March 2022 news article, Yahoo Finance noted that the Ukrainian government and nongovernmental organizations supporting the Ukrainian military effort have collectively raised $59.2 million from crypto donations. Alex Bornyakov, Ukraine's Deputy Minister of Digital Transformation, stated that crypto donations are crucial, especially due to the fast turnaround time: "In times like these, response time is crucial. Crypto is playing a role to give us flexibility to respond really quickly to deliver the army's required supplies." Crypto donations to Ukraine's government began to spike when Mykhalio Federov, Ukraine's vice prime minister, posted a Bitcoin and Ethereum wallet address via his Twitter, soliciting crypto donations worldwide (see **Figure 1**).

Crypto donations like Bitcoins and Ethereum tokens (including NFTs) have helped Ukrainians in a massive way by providing a source of monetary support in a secure fashion, from anywhere in the world to Ukrainians in urgent need. In March 2022, Ukraine government legalized the crypto sector as digital currency donations continue to pour in. It passed a law that creates a legal framework allow foreign and Ukrainian cryptocurrencies exchanges to operate legally. Banks will

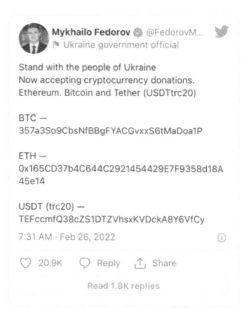

Figure 1   Ukraine's Vice PM Tweeted Crypto Wallet Addresses for Donations

be allowed to open accounts for crypto companies. Although Ukraine did not make any cryptocurrency legal tender, "virtual assets" becomes legal assets.

On crypto-based donations, Tom Robinson, blockchain analytics firm Elliptic's chief scientist, noted in a March 2022 CNBC article that cryptocurrencies also have the advantage of being suited toward international fundraising, due to their decentralized nature: "Cryptocurrency is particularly suited to international fundraising because it doesn't respect national boundaries and it's censorship-resistant – there is no central authority that can block transactions, for example, in response to sanctions."

"No central authority"? Maybe. After the war broke out, US Treasury Secretary Janet Yellen announced that the US would monitor cryptocurrencies as a channel (for Russia) to evade sanctions from the US and Western nations. The International Monetary Fund warned in a report that bitcoin could allow

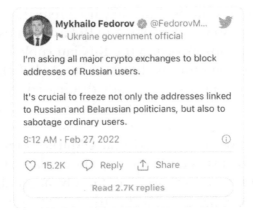

Mykhailo Fedorov @FedorovM...
⚑ Ukraine government official

I'm asking all major crypto exchanges to block addresses of Russian users.

It's crucial to freeze not only the addresses linked to Russian and Belarusian politicians, but also to sabotage ordinary users.

8:12 AM · Feb 27, 2022

♡ 15.2K   Reply   Share

Read 2.7K replies

Figure 2   Ukraine's Vice PM Urged Crypto Exchanges to Sabotage Russian Users

countries such as Russia to monetize energy resources, "some of which cannot be exported due to sanctions." In April 2022, the US Treasury Department began to take action against companies in Russia's virtual currency mining industry, because "these companies help Russia monetize its natural resources." (According to data from Cambridge University, Russia is the world's third-biggest destination for bitcoin mining.)

Decentralized? We will see. (This book will address how "decentralized" the blockchain-based internet can be throughout all chapters.) Whereas Ukraine's vice prime minister tweeted crypto wallet addresses for donations, he also urged crypto exchanges to block the addresses of Russian users. "It's crucial to freeze not only the addresses linked to Russian and Belarusian politicians but also to sabotage ordinary users," he tweeted (see **Figure 2**).

One crypto exchange, DMarket, quickly responded to the calling. According to Axios reports, DMarket is a Ukrainian startup that sells NFTs and virtual items for games such as Counter-Strike: Global Offensive. DMarket soon blocked new user registration from Russian and Belarus on its platform as a manner of protest against Russian invasion, even though approximately 30% of DMarket's customers are from these two countries. Besides the ban on new user registration, DMarket has also frozen assets of Russian and Belarussian users and prohibited transactions involving the Russian ruble.

The crypto community, however, regarded this as a controversial move. Many were greatly displeased with DMarket's actions, expressing via Twitter that Russian users have become the scapegoat caught in the middle of a war they did not cause. Other users have commented that DMarket's move to freeze Russian and Belarussian assets on their platform is stealing value from innocent users who are merely targeted due to their nationality. Furthermore, Twitter users have commented that DMarket's move is a violation of the idea of cryptocurrency and Web3, which heralds decentralization as a core value.

As the war between Russia and Ukraine rages on, it has brought into focus cryptocurrencies and their use. The public attention to the crypto assets is also fired up by the fact that year 2021 has been blockchain and crypto's biggest year ever. Bitcoin and Ethereum hit new all-time highs, NFTs rose from obscurity to front-page news, and we've seen more tech innovation, capital infusion, and institutional buy-in from major companies than ever before. Such crypto debate, however, is merely a tiny part of the confusion around blockchain and Web3, the next generation of the internet.

Terminology in This Book

Web3 advocates suggest blockchain and cryptocurrencies will play a key role in the future of the internet characterized by decentralization. For them, it is a world-changing opportunity to make a better version of the internet and wrest it away from the Big Techs like Facebook and Amazon. At the same time, Web3 has met strong pushback, mostly from voices that are fundamentally skeptical of crypto as a whole. Is Web3 the future, or a scam . . . or both?

The reality is that Web3 is hard to define, blockchain is often confused with bitcoin by many, and the integration of blockchain and Web3 is a nascent idea floated by a mix of buzz, optimism, confusion, theological battles, and pure speculation. As TV show anchor John Oliver put it, these concepts are about "everything you don't understand about money combined with everything you don't understand about computers."

The term Metaverse was first coined in 1992 in *Snow Crash,* a book by Neil Stephenson. Bitcoin, a decentralized peer-to-peer electronic currency, emerged in 2008 as the first and now most popular cryptocurrency. As is the case with so many "tech terms," capitalization is in flux. In this book, we use *Metaverse* as the macro concept virtual-reality-based successor to the internet, and *metaverse* as a general term much like saying "a virtual universe that, in theory, anyone could create." We use *Bitcoin* to describe the concept or entire Bitcoin network, and *bitcoin* to describe the currency.

We put together this book to provide readers a comprehensive and deep dive into how the Web3 metaverse will be generated and built around the world. We dive into use cases that will impact our very existence – art, banking, gaming, payment, trading, music, social media, and more. It aims to provide a detailed guide distilling the complex, fast-moving ideas behind blockchain and Web3 into an easily digestible reference manual, showing what's really going on under the hood.

This book is organized as follows.

Part I: Mega Convergence of Digital Technologies in Metaverse

To support a concept as bold as the Metaverse, we need several orders of magnitude more powerful computing capability, accessible at much lower latencies, across a multitude of devices and screens. Blockchain, the backbone of Web3, is critical for the world awash with data.

Chapter 1: Metaverse – Convergence of Tech and Business Models

This chapter describes that the future of metaverse is built with seven layers of protocol like ISO internet standards, with blockchain technology at the heart of each layer to serve many functions, including governance protocol, incentive mechanism, global payment rail, trustless participation, and global immutable ledger for crucial activities in the metaverse. Value creation and distribution of data are being taken away from centralized actors (like Big Tech companies and sovereign nations) and put into the hands of decentralized groups of individuals.

Chapter 2: Blockchain, the Backbone of Web3

As the decentralized data technology, blockchain will be the foundation of the next generation internet – Web3. This chapter introduces blockchain's four key components: smart contract, public key encryption, consensus algorithm, and peer-to- peer networking. Then it covers the convergence of blockchain with other advanced digital technologies such as Internet of Things (IoT), decentralized storage, AI, cloud computing, and cybersecurity.

Part II: Blockchain Breakthroughs

Just like blockchain is (way) more than bitcoin, Web3 is expanding beyond its financial origins to become the new internet based on ownership and decentralization. The interplay among crypto, DeFi, NFT, gaming, and social work are driving more tech innovation and user cases in the blockchain-based creator economy.

Chapter 3: Cryptocurrencies and Tokenomics

If the growth of crypto as an asset forced everyone to pay attention to the multitrillion market capitalization of the crypto world in 2021, it is the growth of crypto beyond currencies that has the potential to reverberate across industries. This chapter introduces Bitcoin and Ethereum as the origin of cryptocurrencies before diving into more diverse tokens that are emerging in recent years. Diversification means more use cases, and with more use cases comes greater adoption. Because of this positive escalation effect, the crypto industry is branching out.

Chapter 4: DeFi (Decentralized Finance) – Bankless Metaverse

DeFi is the next frontier in finance for the decentralized metaverse. This chapter explains the governance tokens and revenue sources of DeFi, the security issues of layered protocols, and the challenges of DeFi mass adoption. In addition to compete and collaborate with "centralized finance" in traditional finance, DeFi is expanding into completely new territories, such as NFTs, games, and social networks.

Chapter 5: NFTs, Creator Economy, and Open Metaverse

NFTs are playing a major role in bringing blockchain to the mainstream. This chapter explains why NFTs have led to a digital renaissance taking place in the world of art and content creation. While the NFT market started initially with the digital art side, it is going to have broader applications in the creator economy. NFT will become a new tool for consumer businesses, at a time when the line between physical and virtual experiences is blurring. But the NFT market must overcome major challenges to succeed going forward, from speculation to mainstream adoption.

Chapter 6: Blockchain Gaming in Metaverse

This chapter examines how the gaming world, with the new addition of blockchain technologies, has already shown some key elements as to how the metaverse might evolve. Blockchain and digital assets represent the cutting-edge infrastructure level revolution within gaming. The crypto and blockchain technologies are set to disrupt the games and digital entertainment space in a profound way, as gaming content creation and in-game digital assets will broadly move onto the blockchain, as illustrated by play-to-earn (P2E) NFT models.

Chapter 7: Metaverse Privacy – Blockchain vs. Big Tech

Consumer data is going to be at the very heart of the metaverse, and the big tech companies' extensive data gathering in the metaverse will become an even bigger data privacy issue than today. This chapter discusses a few data governance models for metaverse applications, as well as new technologies that can be integrated to provide better privacy protection, including zero-knowledge proof (ZKP), secure multi-party computation, homomorphic encryption, and federated learning.

Chapter 8: Metaverse Security

Blockchain's encryption, immutability, and decentralization attributes make it a great choice for securing data. However, blockchain itself is not free of security risks. As illustrated by the identity security issue and other security topics in this chapter, any metaverse application faces two basic sets of security problems: familiar challenges technologists have been dealing with for decades, and brand new ones built specifically for metaverse settings. This chapter examines the latest cases of security breach in detail.

Part III: Three-Way War among Open Metaverse, Big Tech Walled Gardens, and Sovereign States

Just like in the context of digital currency, where a three-way competition among the cryptocurrencies, Big Tech tokens, and CBDCs (central bank digital currency) intensifies, for Web3 infrastructure, the open metaverse must compete with both Big Tech corporations and government-backed blockchain networks.

Chapter 9: Public Crypto, Government CBDC, and Big Tech Coin

This chapter explains why the crypto ecosystem must fight a currency war on two fronts. On one hand, the war against the financial establishment of governments, including the national CDBCs, which competes with cryptocurrencies for transactions in the Metaverse, and crypto regulations that limits the usage and trading of crypto assets. On the other hand, Big Tech companies such as Meta are trying to provide unprecedented expansion of financial products on a single platform, facilitated by Meta's own coin (Diem) and digital wallet (Novi).

Chapter 10: Web3 Creator Economy on Blockchain

This chapter investigates the opportunities and challenges around the concept of "open metaverse" that operates on open, permissionless, blockchain architecture. The open metaverse is facing a three-way competition with the Big Techs (who are starting to fold Web3 ideas into their centralized platforms), as well as sovereign states that provide permissioned blockchain infrastructure that's cheap and convenient. In the Web3 world, crypto, DeFi, NFTs, and more blockchain-based technologies are converging for a paradigm oriented around the users and their sovereignty: their identity, data, creation, and wealth.

Mega Convergence of Digital Technologies in Metaverse

To support a concept as bold as the Metaverse, we need several orders of magnitude more powerful computing capability, accessible at much lower latencies, across a multitude of devices and screens. Blockchain, the backbone of Web3, is critical for the world awash with data.

CHAPTER 1

Metaverse: Convergence of Tech and Business Models

- Metaverse, Omniverse, and Human Co-Experience
- Big Tech vs. Web3
- Seven Layers of Technology Stack
- Business Models Converging in Metaverse
- Building a Better Internet for the Creator Economy

Metaverse, Omniverse, and Human Co-Experience

In October 2021, Facebook, the company of the world's largest (and beleaguered) social network announced that it would rebrand its corporate identity to "Meta" in order to double down on its commitment on the promise of a "Metaverse." The Metaverse, as Meta describes it, "is a new phase of interconnected virtual experiences using technologies like virtual and augmented reality." Subsequently, the founder Mark Zuckerberg announced, in his new capacity as the Meta CEO, that Instagram will soon enable users to display – and "hopefully" mint – NFTs, the nonfungible tokens on blockchains.

The social network will no longer define the future of Facebook (Meta). The Metaverse will. But what is the *Metaverse*, exactly?

The Metaverse is the convergence of two ideas that have been around for many years: virtual reality and a digital second life. To hear Tech CEOs like Zuckerberg talk about it, the Metaverse is the future of the internet. Or it's about virtual and augmented reality. Or it's a video game. Or maybe it's a deeply immersive version of Zoom (not sure if that would be more uncomfortable)? A virtual world that mirrors our own physical world?

The truth is that the Metaverse may encompass all the above, and it's best understood as the broad term to cover whatever is coming next for the internet. In addition to Facebook, a new generation of major tech, internet, and gaming companies have joined the bandwagon, and they have their own vision for the Metaverse (see **Figure 1.1**). For example:

- **Nvidia**, the Californian chipmaker, instead calls the Metaverse the Omniverse. Its platform is connecting 3D worlds into a shared virtual universe. Omniverse can be used for projects such as creating real-life simulations of buildings and factories. It could be the building blocks of the Metaverse. "We waste a whole bunch of things

Different Tech Leaders' Take on What is the Metaverse

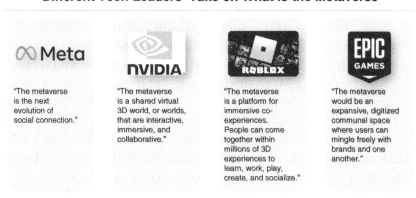

"The metaverse is the next evolution of social connection."

"The metaverse is a shared virtual 3D world, or worlds, that are interactive, immersive, and collaborative."

"The metaverse is a platform for immersive co-experiences. People can come together within millions of 3D experiences to learn, work, play, create, and socialize."

"The metaverse would be an expansive, digitized communal space where users can mingle freely with brands and one another."

Figure 1.1   Big Techs' (Different) Visions for the Metaverse
Source: Official websites of Meta, Nvidia, Roblox. Decrypt.co. *Washington Post*

to overcompensate for the fact that we don't simulate. We want to simulate all factories in Metaverses, in this omniverse," Nvidia CEO Jensen Huang said in an interview with CNBC.

- **Roblox Corporation** is a videogame platform that only went public in the year 2021. But in November the same year, it announced plans for a Metaverse that is built around its players. The company says it wants to create a virtual space where people can "come together within millions of 3D experiences to learn, work, play, create, and socialize." Roblox refers to that as a "human co-experience," a term indicating that the Metaverse is bigger than gaming.
- **Epic Games** has long been in the Metaverse. The company behind the video game Fortnite has become more than just a shooting game. The Epic/Fortnite platform allows gamers to participate in dance parties and virtual concerts, such as one it held for pop star Ariana Grande. "We don't see 'Fortnite' as *the* Metaverse," says an executive of Epic Games, "but as a beautiful corner of the Metaverse."

Meanwhile, the traditional Big Tech companies are also exploring the world of Metaverse, hoping to create new growth opportunities from their existing internet platforms, for example:

- Google made a Metaverse statement with its latest tool Google Lens, which enables users to use a device's camera to capture an object. The technology then uses image recognition and Google's search system to describe what the object is and provide information about it. Such a system could one day be used with headsets in a metaverse.
- Microsoft has started developing a series of "metaverse apps" to help business users of its Azure cloud computing

service combine virtual and physical elements. "Metaverse is essentially about creating games," said its CEO Satya Nadella, "It is about being able to put people, places, things [in] a physics engine and then having all the people, places, things in the physics engine relate to each other."

- Tencent, the social media and gaming giant in China, is reportedly entering the Metaverse, and experts say the virtual world could shape up to be a battle between Meta and Tencent. Tencent has strategic partnerships such as with Epic Games and Roblox's gaming platform. Additionally, Tencent's empire spans virtual offices and mobile payments, so it would have a massive audience across multiple industries.

- Huawei, the 5G and smart hardware leader in China, reached a strategic cooperation with Perfect World, a Chinese cultural and entertainment group, in November 2021. This partnership will integrate metaverse elements into the gaming industry. Perfect World's self-developed ERA engine will work with Huawei's Hongmeng OS (Operating System) to apply distributed computing and shading technology to break the hardware limitation of a single device, potentially providing better game experience.

As we speak, Big Tech companies such as Facebook, Google, Microsoft of the US, and Tencent, Alibaba, and Xiaomi of China are leveraging the convergence of emerging digital technologies, such as the super-fast cellular 5G networks, internet of things (IoT), artificial intelligence (AI), blockchain, cloud computing, and Big Data analytics, AR/VR, game tech, decentralized storage (like IPFS), decentralized and mesh network, and even quantum computing, to create metaverses powered by massive data from both physical and virtual worlds. But an important question arises: Are we keen to migrate into the Metaverse built by Big Tech?

Big Tech vs. Web3

Big Tech gave us the internet as we know it, but that also brought heavy baggage. Just a handful of enormous companies control the web, whether that's Google, Amazon, Microsoft, Apple, or Facebook (and other major "platform companies"), and they're happy to keep it that way. Although the current internet has expanded social connectivity and more user participation (e.g., user-generated content), we have witnessed large-scale walled platforms that require users to operate within the respective app and device. The users are confined to operate within the individual ecosystem of these large platforms.

Equally important, the current internet is dominated by companies that provide services in exchange for your personal data. Because data has become a critical resource in AI and data-driven technologies, the internet giants more often *proactively* collect user data. Because average users are using these popular internet platforms for everything in their daily lives, the internet giants are collecting every aspect of user data, whether identity data, network data, and behavioral data (see **Figure 1.2**). Take precision marketing, for example. Users' data can be analyzed and based on that they are given different characteristic labels (e.g., "keen to travel"; "makeup lover").

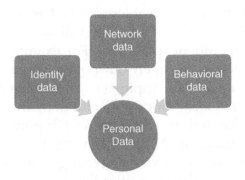

Figure 1.2   Personal Data – Key Resource for the Digital Economy

Then, companies show specific advertising messages to potential customers based on the matching of labels.

- **Identity data.** This includes basic information of a person, such as a name, gender, mobile phone number, and identity card number, which are mainly applied to authenticate users' identities.
- **Network data.** This contains location data, log data, and device information. For example, mobile payment services may encourage users to share location data as well as personal information and purchasing habits with others.
- **Behavior data.** When users browse websites or Apps, those behaviors are recorded to extract user behavioral habits. For example, from the patterns of Facebook "likes," data analysts could predict the users' sexual orientation, religion, alcohol and drug use, relationship status, age, gender, race, political views, and more.

What does this mean for Big Tech companies racing into the Metaverse? The Metaverse can be the next state of the internet's consolidation, a marketing spin on Big Tech's increasing reach and power. Big Tech could re-pitch their extensive lineup of products under a new name, and there would be more data collection from users, since the Metaverse is meant to be a more "immersive" internet. If that's the case, the Metaverse is still a story of Big Tech – just as problem-riddled as now – but bigger.

And even the "new" platforms will not solve the problems of the "incumbent" platforms. For example, Epic Games CEO Tim Sweeney has been outspoken about the threat of a Metaverse run like an Apple ecosystem, governed by "one central company" and "more powerful than any government," he once commented at a VentureBeat report. (Epic Games started a lawsuit against Apple for antitrust violations in 2021, challenging Apple's policy of collecting a 30 percent fee on every in-game transaction in titles like Fortnite. We will cover the case in more detail later in Chapter 10.)

His vision for the Metaverse, shared with *The Washington Post*, entails a cyberspace made interoperable through Fortnite as a game platform and Epic Games' Unreal Engine. Not too surprisingly, the judge on the *Epic v. Apple* case wrote that "Epic Games seeks a systematic change which would result in tremendous monetary gain and wealth . . . [The lawsuit] is a mechanism to challenge the policies and practices of Apple and Google which are an impediment to Mr. Sweeney's vision of the oncoming Metaverse." Ouch.

Similarly, John Riccitiello, CEO of competing game engine company Unity, agrees that Big Tech's vision for the Metaverse is Orwellian. His solution? Everyone should use Unity. "It pulls down the height of the wall of the walled garden," he says. In the history of the internet, things rhyme. From Microsoft in the 1980s to Apple, Google, Facebook and Amazon in the 2010s, all tech giants have started out offering unique services that consumers loved, and they fought for more open competition against incumbents. Over time, after they won leadership positions in the new internet, their missionary zeal waned. They became the "new monopoly."

Therefore, the actual promises of Metaverse, in our view, will not (and should not) be solely realized by Big Tech companies. Who wants a Metaverse built the way Web 2.0 was? (The current mobile internet on smartphones is often referred to as Web 2.0, and the beginning of internet on PC computers is Web 1.0.) Big Tech companies may build up a Web 2.0 Metaverse, as they're not going to give up their server-based models or data collection. And we may even see a much quicker scale of the Web 2.0 Metaverse, building on the existing major platforms.

But new open-source metaverse projects are now seeking to combat the inevitability of this next total service – environment internet by Big Tech platforms. The next web, as the true Metaverse enthusiasts believe, should be architected on open protocols and standards, including blockchain technology. The true spirits of fairness, openness, and community building with the Metaverse ecosystem will come from the decentralized communities on the blockchain. The rapid innovation from open source developers will make their mark on the Web3 Metaverse economy.

Figure 1.3 shows that from an internet user's perspective, Web1.0 is "read only," Web2.0 platform economy is "read and write," and Web3 or the Metaverse will be "read, write, execute, and own." The key is that the ownership of the internet itself should shift from Big Tech companies to individual users. The ultimate vision of Web3 is that there will be no dominant "mega corporation." Instead, the Metaverse will be built by millions of creators, programmers, and designers, each earning a bigger share of the rewards than the tech giants currently allow.

(In this book, Web3 and Metaverse are used interchangeably, and they create a clean break with the present-day internet. Using the Metaverse term as a distinctive descriptor allows us to understand the enormity of that change and, in turn, the opportunity for disruption. And the Web3 term is a direct expression that we need to build a better internet.)

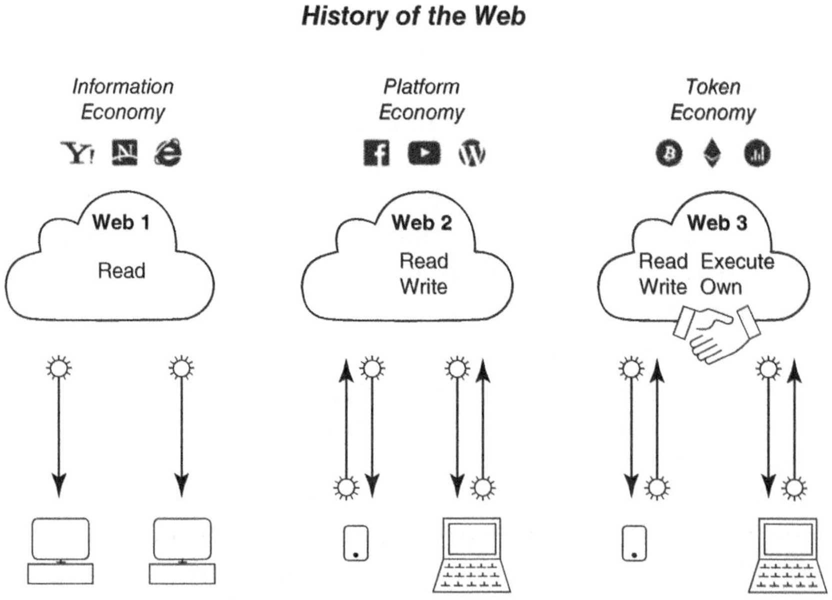

History of the Web

Figure 1.3   From Basic Internet to Web3 (Token Economy)

We argue that the future of Metaverse is built with seven layers of protocol like ISO internet standards, with blockchain technology at the heart of each layer to serve many functions, including governance protocol, incentive mechanism, global payment rail, trustless participation, and global immutable ledger for crucial activities in the Metaverse. In the following sections we will define those layers and describe blockchain's central role in all layers.

Seven Layers of the Technology Stack

Similar to ISO internet standards, the Metaverse internet is composed of seven layers, from the physical and network layer at the bottom (the first layer) to the digital economy of the Metaverse at the seventh layer (see **Figure 1.4**). This section will dive down into each layer of the Metaverse. (For readers who are more interested in the Metaverse business applications, you may skip this section and come back to the technology stack discussion later.)

Layer 1: The Physical and Network Layer

The physical layer includes IoT devices and AR/VR devices. The network layer includes the 5G/6G network and

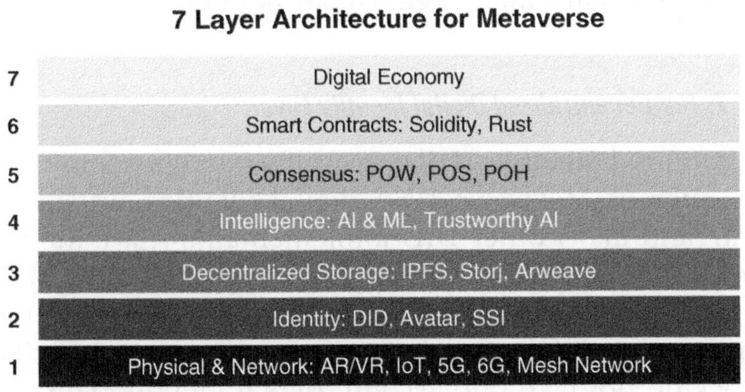

7 Layer Architecture for Metaverse

7	Digital Economy
6	Smart Contracts: Solidity, Rust
5	Consensus: POW, POS, POH
4	Intelligence: AI & ML, Trustworthy AI
3	Decentralized Storage: IPFS, Storj, Arweave
2	Identity: DID, Avatar, SSI
1	Physical & Network: AR/VR, IoT, 5G, 6G, Mesh Network

Figure 1.4   The Seven-Layer Architecture for Metaverse

"mesh network." Because of the speed and pervasiveness of the 5G network, the blockchain transactions can be propagated much more efficiently and network bottlenecks can be reduced and thus improve the performance and scalability of blockchain. The data collected from IoT devices and from AR/VR devices can be propagated to the upper layers with data privacy and sovereignty protected using decentralized identity and then empower the Metaverse economy.

A mesh network allows network nodes to connect directly, dynamically, and nonhierarchically to as many other nodes as possible and cooperate with one another to efficiently route data from/to clients. This lack of dependency on one node allows for every node to participate in the relay of information. Mesh networks dynamically self-organize and self-configure, which can reduce installation overhead.

The ability to self-configure enables dynamic distribution of workloads, particularly in the event a few nodes should fail. This, in turn, contributes to fault-tolerance and reduced maintenance costs. Smart contract can be deployed on top of the mesh networks to incentivize workload execution, bandwidth sharing, and data sharing, which eventually serve as basic building blocks in the Metaverse internet. Blockchain technology can be used to enhance 5G security and enable mesh network connectivity and bandwidth via its immutability property, incentive mechanism, and global payment rail.

Layer 2: The Decentralized Digital Identity Layer

Decentralized identity (DID) or self-sovereign identity solutions, such as Metaverse DNA digital identity Avatar, Serto, Sovrin, and many other DID implementations, are the initial attempt of allowing individuals to manage and control their own identity. The background is that traditional internet designs such as ISO's seven-layer protocol, and four-layer TCP/IP stack do not take into account digital identity. This is one of the main reasons why traditional internet security problems are frequent. That's why digital identities are now being suggested as the new firewall.

We still have time to structure Web3 with digital identity as the underlying core technology and ecological modules. The benefits of introducing DID in the second layer in the Web3 protocol include:

1. **Increasing security** because of decentralized storage of identity data. There is no centralized database for identity, and each user holds and controls its own identity data. Hackers usually have more incentive to hack centralized identity data stores because of the sheer amount of identity data that can be acquired. For DID, the hacker will have less incentive because they have to hack each DID one by one.

2. **Moving authentication** and access control from centralized policy store to end user's wallet application. This increases access control and promotes user awareness of security and privacy.

3. **Enabling KYC/AML** (know your customer/anti-money laundering) with customer consent for metaverse applications. The majority of real-world metaverse applications will need KYC/AML in most countries to meet regulatory requirements. DID can be used to associate verifiable credentials granted from KYC/AML workflow to meet the regulatory requirements. Also, it's a mechanism to enable KYC once and then use everywhere, cutting the cost of regulatory compliance for metaverse applications.

4. **Providing a foundational block** for data ownership authentication, which is critical for the data-sharing economy. In order for metaverse applications to reach their potential, the data must have the right ownership. In Web2.0, data sharing means "copy and paste," and data owners usually lose the ownership of data. In metaverse applications, the data can be shared with an expiration time and the data owner does not lose ownership of the data.

5. **Authenticating off-chain data.** For off-chain data feeder or oracle, if the feeder and oracle are based on **decentralized** identity, the reputation of the oracle or data feeder can be established, and this enables on-chain smart contracts to get accurate data input for business applications.

Layer 3: The Distributed Data Layer

The data will be stored in a distributed and decentralized fashion, using technologies such as IPFS, FileCoin, and BigChainDB. The decentralized peer-to-peer storage system has the following benefits:

1. **Lower costs.** The decentralized data storage system leverages and incentivizes utilization of idle storage, using the token economics model, to reduce the waste of storage and thus reduce overall cost of storage. As decentralized storage markets mature, the overall cost of storage will be much lower compared to the centralized cloud storage system such as AWS S3 or other types of cloud storages. We see decentralized storage gradually taking over the centralized cloud storage market share in the next decade.
2. **Higher reliability.** The data gets distributed and stored on multiple hosts in the decentralized network. The system saves copies of the original data (creating a deliberate data redundancy). In case of any loss or hardware failure, the system will present the backup copy. Additionally, chunks of all shared data can be separately encrypted using a unique hash. This extra security layer protects data from intruders. (The blockchain concept of "hash" will be explained in detail in later chapters.)
3. **Increased speed.** Unlike a centralized storage system, decentralized storage systems use peer-to-peer technology. Data transmissions do not happen through the central

server, which becomes slow at peak traffic times. With advanced routing and load balancing and caching algorithms, in the future, the speed can be improved further. In addition, since several copies of data get stored at multiple locations, downloads can become quicker.

4. **Good price discovery and fair market pricing.** With millions of nodes present, the market for decentralized storage systems becomes a perfect competition. No single node can charge a premium price. This ensures good price discovery and fair pricing across the entire market. Such a market also guarantees that only good-quality nodes can compete and survive.

5. **Increased security and privacy.** Most important of all, decentralized data storage systems provide a high level of security. They partition the data into smaller chunks, make copies of the original data, and then encrypt each portion separately using hashes or public-private keys. The whole process secures the data from bad actors.

Layer 4: The Distributed Intelligence Layer

Artificial intelligence (AI) and machine learning (ML) are currently the core component inside a dynamic Web2.0 tech stack. But the main problem with current AI/ML is their siloed data and proprietary algorithms. The data sharing and algorithm sharing among different organizations introduce privacy nightmare as well as standardization obstacles. In the Web3 and Metaverse era, we see that AI/ML leveraging blockchain technologies become more distributed and decentralized.

By leveraging smart contract and token economy, an incentive mechanism can be provided to AI/ML with high-quality data and algorithms. The AI/ML algorithms' hash can be published on blockchain, such that before each call to AL/ML inside a metaverse application, you can calculate and compare to see if the hash has changed, which can help in determining whether the algorithm has been changed by hackers. You can define a workflow process to vet and publish good-quality data and AI/ML algorithms, using blockchain technologies to sign and execute the workflow tasks.

By implementing role-based access control based on smart contracts, the privacy of data sharing can be managed to allow only authorized users to access the data, so that the privacy concerns can be minimized to allow data sharing among different organizations. The standard application programming interface (API) technology can be used (such as Rest API and GraphQL) to allow standard access to the quality data and algorithms, including incentivized participation from different data providers. (For example, "The Graph" project provides decentralized on-chain data for blockchain projects.)

Layer 5: The Consensus Layer

The consensus layer is composed of one or several hybrid consensus algorithms to make sure that all participants agree on the state of the Metaverse network. From the blockchain technical perspective, a consensus algorithm is a mechanism through which all the peers of the blockchain network can reach a common agreement about the present state of the distributed ledger. In this way, consensus algorithms achieve reliability in the blockchain network and establish trust between unknown peers in a distributed computing environment.

In addition to technical consensus used in the blockchain algorithm, we also see the importance of the so-called "social consensus." In the Metaverse, the "social consensus" means the governance and active participation of individuals or organizations within the Metaverse ecosystem. The social consensus in Metaverse needs to meet the following requirements:

- **Coming to an agreement.** Everyone in the ecosystem strives to reach an agreement, which would benefit the whole Metaverse ecosystem.
- **Collaboration.** Everyone in the ecosystem aims for a better agreement that results in the whole ecosystem's interests.

- **Cooperation.** Everyone in the ecosystem will work as a team and put their own interests aside.
- **Equal rights.** Everyone in the ecosystem has the same right in voting based solely on its stake in the system. The centralization of stake or so-called "whales" in the ecosystem must be dealt with, using technologies such as quadratic voting or other mechanisms.
- **Incentivized participation.** Incentive mechanism needs to be in place to encourage active participation.
- **Borderless.** The social consensus needs to be global and without borders.

Layer 6: The Smart Contract Layer

The smart contract layer can be viewed as an orchestration layer for the Metaverse economy. The complex business logic and workflow process related to critical transactions are executed via smart contracts.

A smart contract, like any contract, establishes the terms of an agreement. But unlike a traditional contract, a smart contract's terms are executed by the codes running on a blockchain like Ethereum, Polkadot, Solona, and HyperLeger Fabric. Smart contracts allow developers to build decentralized apps that take advantage of blockchain security, immutability, integrity, and on-chain verifiability while offering sophisticated peer-to-peer functionality – everything from value exchange, insurance, and loans, to trade finance and gaming. Just like any contract, smart contracts lay out the terms of an agreement or deal. What makes smart contracts "smart," however, is that the terms are established and executed as code running on a blockchain, rather than on paper sitting on a lawyer's desk.

Smart contracts are written in a variety of programming languages (e.g., Solidity, Rust, Java, C++, and Web Assembly). On the public chain ecosystem, each smart contract's code is stored on the blockchain, allowing any interested party to inspect the contract's code and current state to verify its functionality.

Each computer on the network (or "node") stores a copy of all existing smart contracts and their current state alongside the blockchain and transaction data.

Smart contract-powered apps are often referred to as "decentralized applications" or "DApps" – and they include decentralized finance (or DeFi) tech that aims to transform the banking industry. DeFi apps allow cryptocurrency holders from anywhere in the world to engage in complex financial transactions – saving, loans, insurance – that without a bank or other financial institution taking a cut.

In addition to DeFi applications, smart contract will play a crucial role for various decentralized applications in Metaverse, including gaming, education, healthcare, tourism, supply chain management, trade finance, and legal applications, and many more industry sectors. The business flow and associated logic of these industry sectors can be implemented using smart contracts. It's important to note that for real-world metaverse applications, the smart contracts need to get reliable input from both layer 3 (data layer) and layer 4 (intelligence layer) and then leverage layer 5 (consensus layer) and layer 6 (smart contract) to execute the related business logic. The result is vast value creation for the society, thanks to the huge productivity gains.

Layer 7: The Metaverse Economy Layer

The layer 7 is the Metaverse economy layer. The layered architecture allows open platform design and component reusability. The higher layer of protocol can be built on top of the lower layer of protocol. Different products and systems built using the layered architecture can communicate with each other via APIs. (The security issue, however, will be the core of each layer and needs to be applied to each layer with "security first" design principle.)

The Metaverse economy has four core elements: digital creation, digital assets, digital markets, and digital currencies. The first is digital creation, which is the beginning of the Metaverse economy, without which there is no commodity to trade. In the physical world, people "create" all kinds of things or services. We describe it as a *product*; when it enters the market for circulation, it is referred to as a *commodity*. In the Metaverse, people are doing *digital creation* and creating *digital products*. Digital creation is digital and essentially is a collection of data. As the following chapters will illustrate, new digital technologies are now enabling ordinary internet users to become creators of digital contents.

The second is digital assets, which are represented by data and have property rights and can be used in transactions in the Metaverse. The third is the digital market, which represents the digital world marketplaces and the trading rules that everyone must follow. Finally, the fourth is digital currency, which shall enable global digital asset transactions with real-time settlement and clearing at minimum fees. Therefore, the Metaverse economy is essentially the "creator economy."

Business Models Converging in Metaverse

While the Web3 technology stack is still evolving, many companies have joined the bandwagon and announced warm and fuzzy business ideas around metaverse magic. In addition to the Big Tech companies mentioned earlier, major brands like Adidas, Coca-Cola, Dolce & Gabbana, Gucci, NBA, and Nike – just to name a few – also view the Metaverse creating a world of infinite possibilities for them to create new experiences and engage with their customers in entirely new relationship-building ways.

Meanwhile, numerous Web3 startups are emerging in this space to create the Metaverse in the decentralized context. For Big Techs and startups, the common belief is that the Metaverse is the future social network, and more. It will connect everyone and maybe even everything. Thus, the Metaverse provides a platform and ecosystem where business models converge. The case studies in this section will illustrate the convergence of business models at metaverse plays, both from Big Tech and early startups' perspectives.

Case 1: Meta – AI, AR/VR, Big Data, Social Network, and UGC Converging

Facebook has been planning its foray into the Metaverse for some time now – possibly even several years. But renaming the parent company to Meta was perhaps the biggest, boldest statement of intent the firm could make. "The next platform and medium will be an even more immersive and embodied internet where you're in the experience, not just looking at it, and we call this the Metaverse," said CEO Mark Zuckerberg at the announcement of the name change.

Zuckerberg believes that the Metaverse provides an organic development of the company within the existing concept, but there are more reasons why Metaverse makes sense for Facebook (see **Figure 1.5**):

- **Increased engagement.** Virtual reality is supposed to increase the time users spend online and consequently spur content consumption.
- **New content market.** Metaverse offers huge opportunities for creating and selling virtual 3D content, far greater than those of Instagram or TikTok.
- **A new level of communication.** Metaverse will allow people thousands of kilometers away from each other to communicate as if they were sitting in the same room.
- **A new branch of economy.** According to Zuckerberg, the Metaverse must have its own comprehensive economic system.

Meta platforms owns not only four of the top six social media platforms, but also Oculus, which manufactures VR hardware. Virtual reality has been about to go mainstream for a decade now but is far from ubiquitous, leaving the company perpetually trying to capitalize on this $2 billion acquisition. What could sell VR headsets more effectively than the notion

**Meta: the Convergence between AI, AR/VR,
Social Network, UGC, Big Data**

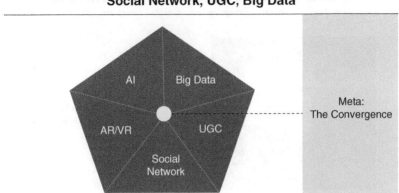

Figure 1.5   Meta – AI, AR/VR, Big Data, Social Network, and UGC Converging

that everybody will need one to access the internet of the future – especially if that same internet is Meta's own?

Case 2: Roblox – 3D Communication, Social Network, AR/VR, and NFT Converging

Human co-experience is a term used by Roblox CEO David Baszucki. In a recent speech, Baszucki said, "It's been called the Metaverse today. We've called it human co-experience." Baszucki defines the Metaverse as a place where technology combines high-fidelity communication with a new way to tell stories, borrowing from mobile gaming and the entertainment industry. According to Baszucki, this new category of the Metaverse or co-experience is predicated on eight fundamentals: identity, social, immersive, low friction, variety, anywhere, economy, and civility.

Essentially, the "human co-experience" can combine business models from 3D communication, social network, and potentially AR/VR and NFT into the Metaverse (see **Figure 1.6**). The users of Roblox can seamlessly shift between modes of communication, from text, to voice, to video, to 3D immersive. In

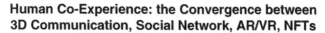

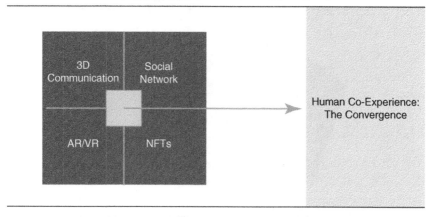

Figure 1.6   Roblox – 3D Communication, Social Network, AR/VR, and NFT Converging

fact, different participants in the conversation may choose the most convenient mode of communication depending on the context. The result is individualized, immersive co-experiences, where people can come together within millions of 3D experiences to learn, work, play, create, and socialize.

Case 3: Soul – Social Network, AI, and Digital Coin Converging

China Tencent-backed Soul App branded itself a "Soul"cial (an extension of "social") Metaverse for the young generation. Soul app went online in November 2016, and it has of late come to overseas markets including North America, Japan, and South Korea.

According to Zhang Lu, CEO of the company, the problem the app took on was that "young people usually have certain emotions and viewpoints that they tend not to share with people around them or on WeChat." The approach was to build an anonymous space that links netizens by their hobbies and values. Soul has tagged itself as a "social networking metaverse," probably to differentiate itself from dating apps such as MOMO and Tantan – both seen as local imitators of Tinder.

The product logic behind Soul is nowhere near as complex: connection and content hold the key to sparking the desire to socialize. For example, Facebook attracts individuals from preexisting social circles, before increasing their engagement using content created by their friends and families. Meanwhile, content-based platforms, like Twitter and Instagram, lure and retain newcomers with their original and captivating content. But Soul has been able to address the challenges of whether to give priority to content over preexisting social circles, or vice versa by applying its distinctive recommendation algorithms at both levels, based on AI and Big Data (see **Figure 1.7**).

Even though Soul boasts a futuristic design and user interface, the product is still far from a metaverse. It is a networking platform built on the online socialization model – not an open world with users "creating content and experiences." Though this app has an avatar customization system, it lacks interaction enabled by human-machine interfaces (HMI) or other AR/VR tools. The platform's monetization mainly relies on VIP subscriptions and e-commerce.

In addition, the token economic system is not fully functioning. The elements that resemble a metaverse are its AI-powered

Soul: the Convergence between Social Network, AI, Digital Coin, UGC, Big Data

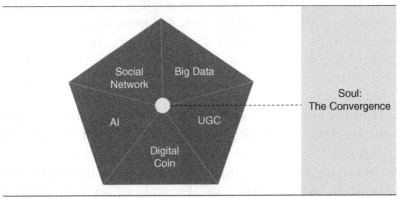

Figure 1.7   Soul – Social Network, AI, and Digital Coin Converging

matching algorithms and a currency called "Soul coin," which is used to purchase avatar decorations and send virtual gifts. Yet, these gifts are not tradable and cannot be exchanged back to "Soul coins." Not to mention that this in-app currency is not decentralized. (The next section will discuss the true Web3 ecosystem properties.)

Like Facebook's Meta, Soul did not make itself a metaverse company by claiming to be one. As for the top-notch game developers backing the project, they simply want to acquire its traffic and data on users' behavioral patterns. True metaverse platforms will be built by decentralized communities, will flourish in the new era of Web3, and will meet the ecosystem properties we define below.

Case 4: Loot – NFT, Creative Ideas, Art, Derivatives, and Games Converging

In late August 2021, Loot, an NFT (nonfungible token) experiment hacked together by Vine co-founder Dom Hofmann, was launched to the public. In the span of a week, the project went viral. Twitter was overflowing with commentary surrounding the project – skeptics, staunch advocates, and everyone in between. Many thought leaders, from Vitalik Buterin to Chris Dixon, framed Loot as a paradigm shift in the conception of the Metaverse, NFTs, and gaming itself. John Palmer went as far as to say, "We're in a different era now; there was "Before Loot and now there's After Loot." Others, however, viewed it as nothing more than a speculative pump in an asset with little intrinsic value." (**Chapter 5** will cover an in-depth discussion of NFTs.)

The concept of Loot was stunningly simple (see **Figure 1.8**). There are 8,000 total "Loot bags," which are text files containing eight phrases. Each of the "items" resembles objects you'd discover in a game like Dungeons & Dragons – that's why Hoffman calls it "adventurer gear." These Loot bags are NFTs on Ethereum that are provably rare, transactable, and composable

Loot: the Convergence between NFTs, Creative Ideas, Art, Derivatives, and Games

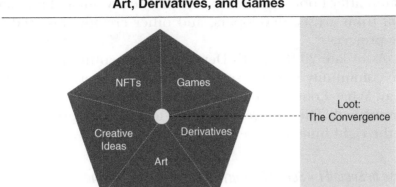

Figure 1.8   Loot – NFTs, Creative Ideas, Art, Derivatives, and Games Converging

with other open protocols. Loot combines the business models of creator economy plus art, derivatives, games, and decentralized autonomous organizations (DAOs) into its own metaverse version and uses the term *Lootverse.*

But what can you actually do with a Loot bag? The answer is very simple and maybe even unsatisfying. According to the project description: "Loot is randomized adventurer gear generated and stored on chain. Stats, images, and other functionality are intentionally omitted for others to interpret. Feel free to use Loot in any way you want."

In other words, there is no game for Loot to be used in, at least in the beginning. The "game," then, is the building process itself; people finding different ways to remix, integrate with, and build on the Loot ecosystem. Loot, then, is a set of open-source objects. Their value comes from the way that they can be used in the future.

Put differently, Loot is a first-of-its-kind bottom-up game. Nobody owns or controls Loot; the original keys to the contract were burnt after a governance vote. Rather, the community of users, builders, and owners determines what Loot means to them and how they want to use their items. In the weeks immediately after Loot's release, there was an early burst of momentum from artists, developers, and other creators inspired by the project.

As of late 2021, Loot's Developer momentum has slowed, the community remains small, and outside interest has faded away. What Loot has achieved already, however, shouldn't be diminished. The Lootverse is likely to become popular again in the right time with the right catalyst.

Case 5: SocialFi – Social Network, Game, Finance, Payment, and NFTs Converging

SocialFi is the convergence of social network, game, finance, payment, and NFT into one platform. The current business model of social networks is inherently extractive. The platforms take their customers' data and sell it, while serving them increasingly intrusive advertising. As the saying goes: Users are not paying for social media; they are the product. Now, SocialFi puts the economics of creation back into the hands of users.

SocialFi aims to deliver benefits and rewards to users through the financialization and tokenization of social influence (see **Figure 1.9**). One such early adopter in SocialFi space is Monaco Planet. By introducing the concept of write-to-earn, content creation itself serves as a form of mining. ("Mining" refers to "earning" crypto tokens on a blockchain ecosystem, which will be explained in detail in **Chapter 3**.) Active content creators and discussion participants on Monaco Planet continuously reap the benefits in the form of native tokens. Most native tokens will be distributed to users who generate content, creating a form of "mining" that is sustainable, inclusive, and productive.

A true SocialFi platform belongs to its users instead of an internet behemoth. And as the vast majority of Monaco Planet's native tokens will be distributed to users as rewards for content creation, Monaco Planet functions as a true decentralized

SocialFi – the Intersection between Social Network, Gaming, Finance, Payments, and NFTs

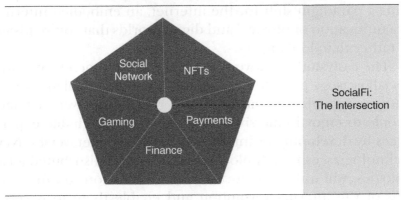

Figure 1.9   SocialFi – Social Network, Game, Finance, Payments, and NFTs Converging

autonomous organization (DAO), governed by native token holders who can send in proposals and vote. As a SocialFi platform, the ownership and governance of Monaco Planet are determined by the users themselves. Moreover, holders of native tokens will enjoy the currency appreciation brought by the platform's growing economic activity.

Building a Better Internet for the Creator Economy

In summary, the years 2021–2022 are the grand opening of Metaverse. The social network giant Facebook rebranded itself as Meta to develop virtual reality digital worlds, the graphics-chip maker Nvidia turned its focus to digital twins – virtual versions of real-world objects or spaces – that people can manipulate and study in computer-generated worlds, and NFT (nonfungible token) rose from obscurity to front-page news, generating digital assets to represent every possible real-world object, from art and music to tacos and toilet paper.

The Metaverse – a persistent, 3D, interactive sequel to today's two-dimensional internet, in which users work, play, buy, and sell inside immersive virtual worlds – has become the internet's Next Big Thing (Web3). As the successor to the mobile internet that has defined the last decade, Web3 represents a paradigm shift for the internet, an embodied internet as a unification of physical and digital worlds that you're inside of rather than looking at.

The case studies above – especially the Monaco Planet and SocialFi cases – provide a glance into what the initial Metaverse business models are and how they could empower ordinary people to enjoy immersive, rewarding, and profitable experiences by developing or interacting in diverse metaverses. New technologies, especially blockchain and other distributed technologies, will unlock opportunity for the billions on the margins of the internet revolution and enable them to become players in the upcoming *creator economy* (see **Figure 1.10**).

By contrast, in the case of Facebook, the Meta rollout has been criticized for user data concerns. Given the track record of Facebook, there are valid reasons to have serious privacy concerns about the company's new focus on virtual reality. Zuckerberg has tried to get out ahead of these concerns, promising multiple layers of privacy protection as the company pivots with its Meta rebrand. The announcement of the Facebook metaverse has thus far been met with at least as much suspicion and hesitancy as it has enthusiasm, as the public wonders what (if anything) the social media giant plans to do differently this time. (Meta stock dropped significantly in the months following the change of corporate name and strategy. But that could be attributable to broad market factors, too.)

Therefore, it's time to build a better internet, where the users, not the Big Tech platforms, control their data value, data privacy, and data security (see **Figure 1.11**). We believe that the next wave of computing innovation—along with entirely new sectors of the economy – will be built on decentralized technology. This is Web3 – a group of technologies that encompasses digital assets, decentralized finance (DeFi), blockchains, smart contracts, tokens, decentralized autonomous organizations (DAOs), and more to come.

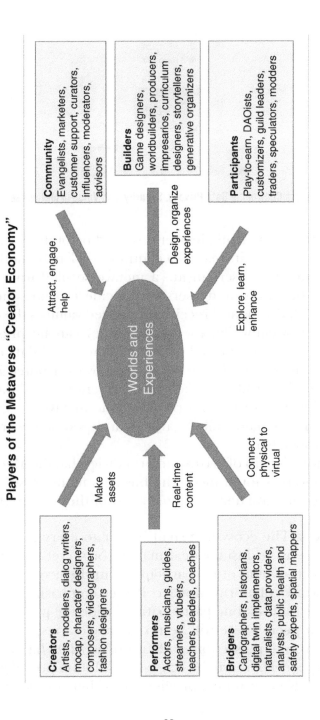

Players of the Metaverse "Creator Economy"

Community
Evangelists, marketers, customer support, curators, influencers, moderators, advisors

Builders
Game designers, worldbuilders, producers, impresarios, curriculum designers, storytellers, generative organizers

Participants
Play-to-earn, DAOists, customizers, guild leaders, traders, speculators, modders

Creators
Artists, modelers, dialog writers, mocap, character designers, composers, videographers, fashion designers

Performers
Actors, musicians, guides, streamers, vtubers, teachers, leaders, coaches

Bridgers
Cartographers, historians, digital twin implementors, naturalists, data providers, analysts, public health and safety experts, spatial mappers

Worlds and Experiences

Attract, engage, help

Design, organize experiences

Explore, learn, enhance

Make assets

Real-time content

Connect physical to virtual

Figure 1.10 The Metaverse "Creator Economy"
Source: Jon Radoff, *Building the Metaverse*

29

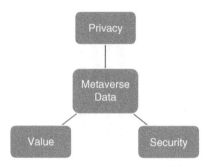

Figure 1.11   Users Controlling Data Privacy, Security, and Value in Metaverse

In our view, the true Metaverse should be built by a decentralized community instead of a centralized IT behemoth. The data generated by users inside a metaverse platform should belong to the users, and users can decide to share the data with other platforms and receive monetized value as the owner of the data. Meanwhile, user data privacy can be preserved since there is no centralized collection, and users' data can be stored on metaverse platforms powered by distributed ledger technology (DLT) like blockchain. (Currently, users' data are "trapped" at Big Tech platforms, and the metaverses created by them may remain walled gardens and potentially create a dystopian society.)

Following is a list of ecosystem properties that the Metaverse business model should meet; the business model cases in the earlier section have demonstrated a few of these properties:

Fairness. The ecosystem shall be fair to every participant, there is no insider deal making or secret transactions which exploit other participants in the ecosystem. The SocialFi platform Monaco Planet emphasizes the fairness for everyone participating inside the platform.

Peer to peer. There will be no intermediaries; all interactions and transactions and various activities happen peer to peer. There will be ecosystem solution providers who are also consumers or clients of other solutions inside the ecosystem. Almost all blockchain-based projects developed by decentralized communities have tried to make peer-to-peer interaction/communication/transaction as the central component of the platform, completely opposite to tech conglomerates (like Facebook) and their metaverse plays (like Tencent's Soul).

Global payment rail. Must have global real-time instantaneous settlement and clearing for payment enabled by cryptocurrency.

Decentralized autonomous organization (DAO). DAO will manage and govern the business relationship, transactions, and activities.

Sustainable token economy. Combining physical and digital worlds with a sustainable business model and actual value creation, Metaverse will be able to flourish and benefit all participants. The token economics must encourage and incentivize participation and contribution from ecosystem players, rewarding positive contribution and punishing malicious actions. Ponzi scheme-like systems that use later arrival participants' funds to pay for early participants can only survive a short duration of time and will not be sustainable.

Security. Security shall be the most important aspect of the Metaverse platform. A defensive, in-depth approach must be implemented to protect every technological layer and build blocks in the Metaverse ecosystem. In addition to cyber and technical security, the ecosystem must consider token economy security and regulatory compliance.

Self-sovereign identity (SSI). SSI means that individuals should own and control their identity without the intervening third party and centralized authorities. Personal data is stored and managed in a decentralized manner, thus increasing its protection. Owners have access to the information associated with their identity and must provide consent before it can be shared. SSI is the foundational building block for the creator economy, as users will all become creators of digital assets.

Immersive experience. An "immersive experience" allows a person to enjoy a more engaging, rich, and rewarding experience than from today's two-dimensional screen. Immersive technologies create distinct experiences by merging the physical world with a digital or simulated reality. Augmented reality (AR) and virtual reality (VR) are two principal types of immersive technologies.

Multiple-dimension experience. The internet experience can be enhanced by multiple spatial and time dimensions, which allow users to teleport to different 2D or 3D spaces in milliseconds and time travel to the past or to the future.

In the near future, blockchain-based Web3 will surround us, with our lives, labor, and leisure all taking place inside it. The *blockchain internet* is poised to revolutionize every industry and function, from finance and healthcare to media entertainment and real estate, creating trillions in new value – and the radical reshaping of society. In the next chapters, we will discuss the convergence of digital technologies that will enable metaverse applications, and we will also introduce different applications and security and privacy aspects of the Web3 Metaverse.

The Metaverse may be the next major platform in computing after the world wide web (Web1.0) and mobile internet (Web2.0). It will represent a profound shift in the way individuals and communities use technology. Value creation and distribution of data is being taken away from centralized actors and put into the hands of decentralized groups of individuals. In addition to Big Tech companies, sovereign nations are aggressively investing into Metaverse research and next-generation digital infrastructure, including state-backed blockchain networks. The race to build the new, decentralized, blockchain-powered internet – otherwise known as Web3 – is on.

2

Blockchain, the Backbone of Web3

- Basic Blockchain Concepts
- Blockchain's Four Key Components
- Mega Convergence of Data Technologies
- Blockchain and Cloud Computing
- Blockchain and Cybersecurity
- Five Challenges of Blockchain Adoption and Possible Solutions
- Why Blockchain Is Essential for Metaverse

Basic Blockchain Concepts

The internet and its digital revolution have one over riding problem: digital data can be replicated. In essence that means that any digital system can be compromised. Protecting the integrity of the digital world, therefore, has proven to be a monumental task. As the Metaverse intends to connect everyone and potentially even everything digitally, at the heart of the Metaverse economy (and society) is the explosion of insight, intelligence, and information – data. As a result, the data management issue is more critical than ever for the new internet.

That's why blockchain is the backbone of the Metaverse. Before we go into specific functions of blockchain applications and analyze how blockchain and cutting-edge digital technologies – such as artificial intelligence (AI), virtual and augmented reality (VR/AR), and Internet of Things (IoT) – may converge to power the Metaverse, let us start with the basic blockchain concepts.

So, what is blockchain?

Using the official definition by NIST (National Institute of Standards and Technology):

> Blockchains are distributed digital ledgers of cryptographically signed transactions that are grouped into blocks. Each block is cryptographically linked to the previous one (making it tamper evident) after validation and undergoing a consensus decision. As new blocks are added, older blocks become more difficult to modify (creating tamper resistance). New blocks are replicated across copies of the ledger within the network, and any conflicts are resolved automatically using established rules.

Wikipedia has a similar definition:

> A blockchain is a growing list of records, called blocks, that are linked together using cryptography. Each block contains a cryptographic hash of the previous block, a timestamp, and transaction data (generally represented as a Merkle tree). The timestamp proves that the transaction data existed when the block was published in order to get into its hash. As blocks each contain information about the block previous to it, they form a chain, with each additional block reinforcing the ones before it. Therefore, blockchains are resistant to modification of their data because once recorded, the data in any given block cannot be altered retroactively without altering all subsequent blocks.

In short, blockchain is a distributed ledger, or database, shared across a public or private computing network. Each computer node in the network holds a copy of the ledger, so there is no single point of failure. Every piece of information is mathematically encrypted and added as a new "block" to the chain of historical records. Various consensus protocols are used to validate a new block with other participants before it can be added to the chain. This prevents fraud or double spending without requiring a central authority.

In addition to the concept of "distributed digital ledgers" or "list of records," the other key tenants of blockchain technology include smart contract, public key encryption, consensus algorithm, and peer-to-peer networking. These technologies work together to enable various foundational infrastructure and applications in blockchain. Blockchain contains many different technologies, including game theory, time stamp, transaction ordering, and distributed computing. In the following section, we will discuss blockchain's four key components in detail (see **Figure 2.1**).

Blockchain's Four Key Components

Smart Contract

NIST defined the smart contract this way:

> A collection of code and data (sometimes referred to as functions and state) that is deployed using cryptographically signed transactions on the blockchain network. The smart contract is executed by nodes within the blockchain network; all nodes must derive the same results for the execution, and the results of execution are recorded on the blockchain.

Smart contract was initially introduced in the Ethereum blockchain and now is widely enabled in many different blockchain ecosystems, such as Avalanche, Binance Smart

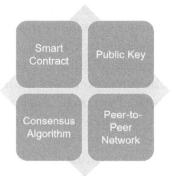

Figure 2.1   Blockchain's Four Key Components

Chain (renamed BNB chain in February 2022), Polygon, and Polkadot. Smart contract programming logic demonstrated its power in early 2016 with many initial coin offering (ICO) and recently has been used widely in DeFi, NFT, GameFi, and metaverse applications. Without smart contract programming, the blockchain technology may only be used in a small-scale peer-to-peer payment application.

Public Key (Encryption)

Public key encryption is also called asymmetric key encryption. As the name suggests, two asymmetric keys (i.e., two different keys) are used for public-key encryption. One key is used for the encryption process, and another key is used for the decryption process. See **Figure 2.2** for a high-level illustration of the public key encryption used in a bitcoin transaction. (Bitcoin is the first and arguably most successful decentralized digital currency to have gained adoption in the world. Users can send or receive payments in bitcoin through a peer-to-peer (P2P) network, which is supported by its underlying blockchain protocol. Detailed discussion in **Chapter 3**.)

In this diagram, Bob wants to send Alice a bitcoin. He uses a bitcoin wallet. The bitcoin wallet will use Alice's public key to

Bob sends Alice bitcoin

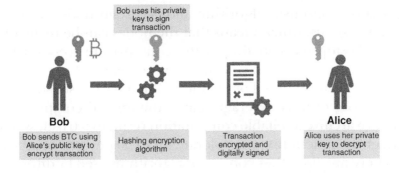

Figure 2.2   Public Key Encryption in Bitcoin Transaction

encrypt the transaction so no one else can unlock the bitcoin inside the transaction. Bob also has to sign the transaction with his private key. This is to prove to Alice and the whole network that he has the ownership of this bitcoin. The bitcoin was broadcasted in a P2P network and validated by network nodes called "miner node," who use Bob's public key to validate Bob's signature. If all went well, there is a consensus, and the transaction is recorded in the blockchain network. Alice is the only one to unlock that bitcoin since the transaction is encrypted by Alice's public key. Eventually, Alice will use her private key to unlock the transaction and spend the bitcoin. (The concept of "miner" and "mining" will be explained in Chapter 3.)

Consensus Algorithm

Consensus algorithm in its simplest definition is for different and distributed computers to agree on the same state of the blockchain and archive the agreement without putting any trust on any participating computers or the "nodes" in the blockchain network. The carefully designed consensus algorithm must consider the unreliable and asynchronous nature of peer to peer networks.

A good consensus algorithm allows certain nodes to be in the faulty status, which means that the nodes can be malicious or simply stop participating in the consensus process yet the blockchain network can still reach agreement with the majority of the honest node. For example, Bitcoin's consensus algorithm proof of work leverages game theory and computation power to archive a reliable computation result in a trustless network environment that can tolerate up to 50 percent of attack.

There are also other consensus algorithms used by different blockchain networks such as proof of stake, proof of elapsed time, proof of history, and proof of replication. The key idea of these consensus algorithms is to allow blockchain nodes to agree on the blockchain transactions and the state of ledger in a trustless and distributed manner without putting too many assumptions and constraints on the underlying network. The fundamental assumption of any consensus algorithm is that the majority of the network nodes are honest nodes and, as such, the network can allow some "bad apples" or malicious nodes in the network.

The consensus algorithm can be extended from the pure technical algorithm to social behavior in the blockchain ecosystem, which is referred to as *social consensus*. A blockchain ecosystem is not just composed of lifeless computer nodes; it also needs human participants, developers, investors, buyers, players, creators, wallet holders, exchange markets, and even government entities and regulators. The "social consensus" means the agreements among the individuals, and organizations on the value of the blockchain network, the fairness of the blockchain enabled token economy, the willingness of participation, and contribution to the blockchain network.

The social consensus is implied in the DAO (decentralized autonomous organization) structure to prompt the common interest and governance of the blockchain network. Both technical consensus and social consensus will play significant roles in metaverse applications. Because the "social consensus" of DAO creates a new decentralized governance model, DAO may replace corporations as the popular organization structure in the Metaverse, which will be discussed in Chapter 10.

Peer-to-Peer (P2P) Networking

Although it is not a new concept and has existed long before the Bitcoin network existed, P2P network is a departure from the traditional client/server architecture where there are separate computers to act as servers to serve different sets of computers as clients. In a P2P network, each computing node can be both server and client. The peer node can discover a set of other peer nodes to communicate without any authentication or authorization. This can allow the information to flow freely and allow the network to be more resilient to a potential system crash.

In blockchain, the peer node can broadcast the block, participating in consensus algorithms, validating each peer-to-peer transaction, and getting rewards for its contribution. The idea of P2P network can be extended into the blockchain to advocate the idea that each individual or organization in the blockchain network has an equal voice and equal right. Everyone's power can be somehow measured by its stake (from the proof of stake (POS) consensus algorithm's perspective) in the network or the electricity power they can use for the network (from the proof of work (POW) consensus algorithm's perspective). No individuals are above any other if they all have equal stake in the network. (POW and POS will be discussed relating to bitcoin in **Chapter 3**.)

Mega Convergence of Data Technologies

As the decentralized data technology, blockchain will be the foundation of the next-generation internet – Web3. As mentioned, the web as we know it today was developed from static content publishing Web1.0 to the current centralized Web2.0.

Web2.0 integrates SoLoMoCo (Social, Location Based, Mobile App, and Cloud Computing) technologies, and it has generated tremendous economic benefits, resulting in big internet companies like Amazon, Google, Facebook, Alibaba, and Tencent.

Web3 is the mega-convergence of more technologies. To support a concept as bold as the Metaverse, we need several orders of magnitude more powerful computing capability, accessible at much lower latencies, across a multitude of device and screen. In addition to SoLoMoCo, we see that Web3 is empowering Metaverse further by converging AI, blockchain, Big Data, decentralized identity (avatars), fintech, decentralized storage, IoT, AR/VR, video rendering, game tech, quantum computing, and more.

More than ever, data will explode in the Metaverse. Humanity's current rate of data creation has us doubling the world's data every two years, and this pace is expected to increase. By 2025, the amount of data will double every 12 hours – or less. This wealth of data, created exponentially as we go about our lives, has the potential to change the ways we live, work, and invest – but only if we have the digital technologies like blockchain to manage it, secure it, and use it with proper privacy protection. (See **Box: Blockchain and Digital Transformation.**)

Meanwhile, the COVID-19 pandemic has highlighted the inadequacies of the world's collective approach to data. The inability, and sometimes unwillingness, to share and use data to combat COVID-19 or to protect against predatory uses of data has negatively impacted individuals, private enterprise, citizens, research institutions, and governments around the globe. A lack of trust combined with asymmetric economic interests are slowing progress – especially in the cross-border context. The explosion of ransomware and software supply chain attack in year 2021 exposed the vulnerability of the centralized data storage within enterprises, big and small.

Blockchain and Digital Transformation

The era of digital transformation is here for companies of all sizes and types, from Fortune 500s to startups. The term refers to companies leveraging enhanced technology to improve their business capabilities, operational efficiencies, and ultimately, their customers' experiences. One important goal for digital transformation is to leverage accurate and trusted data to digitized business processes.

Blockchain technology can be used to assist digital transformation by bridging the gap between the digital and physical worlds. To implement a data strategy between the two worlds, the corporation must ensure two things: data authenticity and data usability.

Data authenticity refers to how data is generated in the physical world and how it is uploaded and verified. It is difficult to achieve authenticity through software alone. An effective way is to use a hardware chip to "sign off" the data with a private key embedded in the chip. Since a digital signature created by the private key cannot be forged and tempered with, the signature can be verified on the blockchain node using its corresponding public key.

In addition, during data transmission, the encryption can be applied to ensure data privacy and data integrity. Blockchain technology can be used to validate hardware signatures, enable P2P and safe transmission protocols, and perform on-chain verification to ensure that the entire process from off-chain to on-chain is safe – that is, confirm that the data is authentic.

Data usability refers to the existence of actual usable data among a large amount of noisy data in the physical world. Different data has different value, and only highly relevant and usable data has high values. Thanks to the blockchain's incentive mechanisms and smart contracts, the frequency of data being used by smart contracts can be used to price data. When data is priced by how data is used, the data price also indirectly reflects the data's effectiveness.

A proof of contribution (PoC) consensus algorithm can be designed to determine the quality and usage of the data, so as to make the digital transformation process more objective and effective. When accurate, usable, and effective data is obtained and analyzed, AI/ML algorithms can analyze the data and store the output of their computation onto the blockchain, enabling applications such as real-time regulation, fraud detection, and credit monitoring in the Metaverse platform.

Furthermore, blockchain technology can improve the accuracy of data, too. Before the data can be stored on a chain, the data is cross-checked and examined by the blockchain network's participating nodes via consensus algorithm, and the data must be digitally signed to confirm the validity and ownership of the data. This improves the accuracy of data.

The current crisis illustrates that without proper technology, protocols, and governance, society risks creating either a world in which access to data is overly restricted and impedes human progress and innovation, or one in which data-sharing solutions are created without properly respecting the rights of the individual parties involved, including businesses. That's why blockchain is critical for the Metaverse.

Of course, blockchain alone has limited use cases if it is not integrated with other technologies. They can work together to resolve the problems of data authenticity for off-chain data, cybersecurity, risk control, and data governance in Metaverse. The different technologies can feed into each other and create an ecosystem of automation – IoT devices collect data on millions of devices, which data is distributed stored, managed by blockchain, and then collated in the cloud and used to train and improve AI algorithms for real-life applications (see **Figure 2.3**).

As these technologies interact and improve each other, the huge synergies will spur more innovation. In the next few sections, we will discuss the convergence of blockchain with such technologies as Internet of Things (IoT), decentralized storage, AI, cloud computing, and cybersecurity. (The latter two will be discussed in their respective sections.)

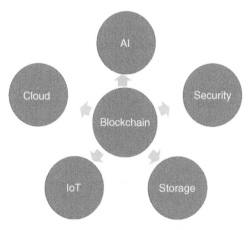

Figure 2.3   Blockchain Converging with AI, Cloud, and IOT

Blockchain and Internet of Things (IoT)

While IoT focuses on improving the collection of appropriate data used for various purposes, blockchain focuses on ensuring that data integrity stays intact. IoT allows devices to transfer data to blockchain networks, creating tamper-proof shared transaction records. Business partners may exchange and access IoT data using blockchain, which eliminates the need for central control and management. In addition, all the transactions can be verified to minimize disputes and to build trust among the network participants.

Using blockchain for IoT has several benefits:

- Reduce the risk of tampering with data received from the IoT devices and thus enable trust among parties in IoT business.
- Leverage blockchain's consensus algorithm for data verification and integrity checking and reduce the overhead of intermediaries.
- Create an immutable audit trail to avoid repudiation risk.
- Enable machine-to-machine payment using smart contract and token economy.
- As edge computing becomes important for IoT, blockchain technology can be used to incentivize participants of edge computing to provide high-quality data while processing and verification on the edge before uploading to the blockchain ledger.

Freight transportation is an application example of IoT and blockchain converging. Moving freight is a complicated procedure that involves multiple parties with different priorities. Temperatures, position, arrival times, and status of shipping containers can all be stored on an IoT-enabled blockchain as they travel. Immutable blockchain transactions ensure that all parties can trust the data and act swiftly and efficiently to carry out their activities.

Blockchain and Distributed Storage

For the future of blockchain development, storage is an indispensable function. Many application scenarios such as AI and IoT require a large amount of data storage, and the integration of these technologies with the blockchain will inevitably need to consume lots of data storage, which is distributed and always available. At present, mainstream blockchains cannot directly store large-scale amounts of data, because full nodes need to synchronize all blockchain data. If a large amount of data exists on the chain, the node load will be too large and the efficiency of the blockchain will become low. Therefore, the most popular way to put data on the chain is to place the data in decentralized storage such as IPFS, while storing small amounts of data such as hashes, meta data, and transaction logs on the blockchain.

However, IPFS technology alone cannot encourage participating network nodes to store the off-chain data and make them available all the time. In order to implement the incentive mechanism, the Filecoin project proposed proof of replication and proof of space time. Proof of replication and proof of space time ensure that the data can actually be replicated, stored, and available for retrieval all the time by miners through incentive mechanisms. We can further use decentralized storage technology such as IPFS and FileCoin to improve data availability. However, the industry still needs to verify whether this mechanism is mature and stable.

Blockchain and Artificial Intelligence (AI) and Machine Learning (ML)

AI/ML can help blockchain become smarter in its real-world applications. For example, in supply chain finance, in order to control risks, AI/ML can be used to provide the blockchain with data processed by AI/ML. Furthermore, AI/ML can analyze abnormal behavior and provide fraud detection for on-chain transactions to alert chain operators of malicious activities. Compared with traditional AI applications, when combined with blockchain technology, the consensus mechanism ensures that the AI/ML training data and algorithm can be verified, and the hash of the AI/ML algorithm can be stored and verified on the chain to ensure both the accuracy of the AI/ML algorithm and output of such algorithms.

Blockchain can also help AI improve data-sharing capabilities by leveraging smart contract-defined incentive mechanism. To mitigate privacy concerns, zero-knowledge proof (ZKP), secure multiparty computing (MPC), and homomorphic encryption algorithm can be applied. Through the use of smart contracts and incentive mechanisms, one can implement data rights verification and data exchange transactions, as well as data pricing.

Furthermore, blockchain can help AI promote optimized AI algorithms, establish distributed computing capabilities for deep learning tasks, and effectively use idle computing resources. Blockchain smart contracts can be used to manage the behavior of AI algorithms and avoid security problems caused by improper use of AI algorithms.

Blockchain and Cloud Computing

Cloud technology (also known as *cloud computing*) means that servers, data storage, databases, networking, software, and analytics are hosted on the internet and stored on large, privately owned data centers. Cloud computing platforms like AWS (Amazon Web Services), Microsoft Azure, and Google Cloud Platform provide three types of cloud offerings:

1. **Infrastructure as a Service (IaaS).** IaaS cloud offering includes computing, networking, and storage services to cloud consumers.
2. **Platform as a Service (PaaS).** PaaS providers offer middleware, database, runtime environment, and development tools.
3. **Software as a Service (SaaS).** SaaS provides actual implemented applications to the cloud consumers so cloud consumers do not need to install any software or purchase computing resources such as servers, storage, and network equipment to use the applications on the cloud.

The cloud provider enables end users to "rent" and remotely access IT resources from cloud providers on a pay-per-use basis, which provides an efficient alternative to the local hosting and operation of IT resources. Businesses subscribe to the cloud services and pay either a monthly or annual fee, just like buying electricity from a power grid. This fee is determined by the amount of data and number of users a business requires – making it easier for a company to scale its operations up and down. Overall, cloud computing provides both improved system scalability and cost savings over traditional IT infrastructure.

In summary, cloud computing offers many advantages compared with traditional data centers hosted by individual companies, which is critical for metaverse players that must handle explosive growth of data:

- It eliminates upfront investment in costly IT infrastructure.
- Cloud computing providers offer pricing models like "pay-as-you-go," which means that consumers only pay for what they use.
- Users don't need to work on provisioning and managing IT infrastructure, since the cloud provider handles that.

Nevertheless, cloud computing providers can suffer from service outages and usually have more negative implications than the outage experienced by traditional data centers. For example, on December 7, 2021, Amazon's web-hosting platform suffered a major outage, taking down major websites such as Facebook, Netflix, Disney+, and Venmo. Also, Amazon delivery service was disrupted, preventing drivers from getting routes or packages and shutting down communication between Amazon and the thousands of drivers it relies on.

The outage also took down several cryptocurrency exchanges such as Coinbase, Binance, and a number of other blockchain-related services that hosted their application on Amazon. Even the decentralized derivatives exchange dYdX was affected, raising questions on how much decentralization of a DEX (decentralized exchange) there really is if the front-end user interface (UI) is hosted at a centralized cloud.

Leveraging IPFS (InterPlanetary File System) to host front-end UI for Web3 applications may become a trend, as metaverse applications seek to get protection from cloud service outage. Unlike HTTP protocol (the existing internet) that locates objects (text files, pics, videos) by which server they are stored on using URL, IPFS locates objects by the hash on the file. IPFS creates file hash per the content in the file. So, if you want to access a particular page, IPFS will ask the entire network if anyone has the file that corresponds to this hash, and a node on IPFS that has this hash will return the file, allowing you to access it.

As such, IPFS uses *content addressing* at the HTTP layer. This means that the content can be used to determine the address of the content. The mechanism is to take a file and hash it cryptographically, which will give you a secure representation of the file. This ensures that no one else can just come up with another file that has the same hash and use that as the address. In addition, all files are distributed and replicated globally, which can be permanently stored via an incentive mechanism powered by blockchain technology.

This is just one example of using blockchain to fix the problem of cloud service outages. In addition, blockchain can be designed to leverage idle computing resources in a decentralized fashion. There are many idle computing resources such as CPU, GPU, storage, and network bandwidth on the planet, but they cannot be utilized by current cloud providers due to their centralized nature. A better approach to cloud computing is to leverage blockchain technology to incentivize computing resources, storage, and network bandwidth sharing, while reducing the risk of security breaches at the same time.

As examples, we will discuss how to leverage blockchain at IaaS, PaaS, and SaaS offerings, respectively in the remainder of this section.

IaaS Offering

For IaaS offering, we can look at examples for (1) compute power or central processing unit (CPU) sharing; (2) storage sharing; and (3) network bandwidth sharing.

1. **CPU sharing.** One example is SETI@home, which is a scientific experiment, based at UC Berkeley, which uses internet-connected computers in the search for extraterrestrial intelligence (SETI). You can participate by running a free program that downloads and analyzes radio telescope data. The program is now in hibernation. One of the main reasons why this excellent program has stopped is due to lack of incentive mechanism and reward scheme for people who have free CPU cycles to participate.

    The blockchain projects such as Three Fold and IEx are trying to use token economy to create peer-to-peer computing power sharing in "decentralized cloud." It aims to build a P2P public overlay network to connect everything on the planet. Connections are end-to-end encrypted and take the shortest path. The team also intends to use token economy to incentivize the network sharing.

2. **Storage sharing.** The best example is FileCoin. Filecoin is an open-source, public cryptocurrency and digital payment system intended to be a blockchain-based cooperative digital storage and data retrieval method. It is made by Protocol Labs and builds on top of InterPlanetary File System (IPFS), allowing users to rent unused hard-drive space. According to Filecoin's authors, it is a decentralized storage system that aims to "store humanity's most important information." Filecoin is an open protocol and backed by a blockchain that records commitments made by the network's participants, with transactions made using FIL, the blockchain's native currency.

    Contrary to centralized storage methodology, Filecoin aims to store data in a decentralized manner, which is resistant to problems that occur in centralized storage. Due to Filecoin's decentralized nature, it protects the integrity of data's location, making it easily retrievable and hard to censor. It also allows people on their network to be their own custodians of the data that they store. Additionally, Filecoin rewards the network nodes that mine and store data on their blockchain network. Similar blockchain-based storage sharing systems include Sia, MaidSafe, and Three Fold.

3. **Network bandwidth sharing.** For example, Helium is a global, distributed network of hotspots that create public, long-range wireless coverage for LoRaWAN-enabled IoT devices. Hotspots produce and are compensated in HNT, the native cryptocurrency of the Helium blockchain. The Helium blockchain is a new, open source, public blockchain created entirely to incentivize the creation of physical, decentralized wireless networks. Today, the Helium blockchain, and its hundreds of thousands of hotspots, provide access to the largest LoRaWAN network in the world.

PaaS Offering

For PaaS offering, the best example would be the various metaverse, DeFi, and NFT applications that serve as "metaverse Legos," meaning that they can be reused to develop new metaverse applications. For development tools, Remix is a good example of a PaaS tool for developing smart contract solidity code. Another good example is Infura API service, which allows developers to interact with Ethereum or IPFS service. Infura is one of the most widely recognized PaaS tools among developers for connecting to Ethereum and IPFS, and it already handles billions of API requests per day.

SaaS Offering

There are many successful examples of SaaS offering. Among them, the best known are "The Graph" and "Chainlink" projects.

With DeFi and NFT (discussed in the next two chapters), the Dapp developed and deployed on blockchain can all be viewed as SaaS offering and supported by decentralized nodes.

- **The Graph Network** decentralizes the GraphQL API and query layer of the internet application stack. Developers can run a Graph Node on their own infrastructure, or they can build on The Graph hosted service. In The Graph Network, any Indexer will be able to stake Graph Tokens (GRT) to participate in the network and earn rewards for indexing subgraphs and fees for serving queries on those subgraphs. Consumers will be able to query this diverse set of indexers by paying for their metered usage, providing a model where the laws of supply and demand sustain the services provided by the protocol.
- **Chainlink** is a decentralized blockchain oracle network built on Ethereum. The network is intended to be used to facilitate the transfer of tamper-proof data from off-chain sources to on-chain smart contracts. Its creators claim it can be used to verify whether the parameters of a smart contract are met, in a manner independent from any of the contract's stakeholders by "connecting the contract directly to real-world data, events, payments, and other inputs."

    Chainlink is operated with a SaaS model. The consumer of Chainlink pays $Link, the platform token of Chainlink, to receive tamper-proof data from this platform. The Chainlink nodes provide trusted data to earn the $Link in a trustless fashion, whereas the node that provides inaccurate data can be punished with $Link.

The decentralized cloud computing technology on blockchain is currently only in its infancy, but companies such as Google, Microsoft, Amazon, and IBM have already begun to conduct research in these areas. Although the specific details have not been made public, one can imagine that cloud computing in the future can be more secure, stable, efficient, energy saving, and more personal computing resources can be effectively utilized. The business model of cloud computing and the operating model of providers are evolving rapidly because of the empowerment of blockchain.

Blockchain and Cybersecurity

The immutability of the blockchain is its important security property, and it effectively guarantees the integrity of the data. However, when integrating blockchain into metaverse applications, we need to adopt a defense-in-depth approach. Some of the top security controls for blockchain include (but are not limited to) smart contract security, consensus algorithm security, blockchain node hardening, crypto exchange security, identity and access management, node-to-node traffic encryption, and on-chain and off-chain data encryption.

Smart Contract Security

Smart contracts cannot be modified once they are deployed on the blockchain *mainnet* (a term used to describe when a blockchain protocol is fully developed and deployed), so if there are security vulnerabilities, they will often cause direct economic losses and are difficult to recover. A classic example is the attack on the DAO.

The DAO, or decentralized autonomous organization, is a program that was built on the Ethereum blockchain network that aimed to become the largest crowdsourcing platform. DAOs aimed to replace management structures that were centralized by using a technologically democratic approach, where decisions are taken by investors and stakeholders. After the DAO smart contract deployed on the Ethereum mainnet, one hacker spotted a flaw in the DAO's code and managed to steal about $50 million worth of Ether, which sent the Ethereum community into panic mode and eventually caused the hard fork of Ethereum.

Another recent smart contract attack example is Poly Network attack. Poly Network is a cross-chain network that essentially allows two or more blockchains to "communicate with each other." To be more precise, it enables users to make transactions across different blockchains without having to convert the digital coins in an exchange. This China-based platform specifically sits on top of several blockchains, including Bitcoin, Ethereum, Binance Smart Chain, Neo, and Elrond.

On August 10, 2021, Poly Network reported that a group of attackers had hacked a smart contract of its network, transferring roughly $610 million (mostly in Ether, Binance Coin, and USDC) and moving them to external wallet addresses. According to the cybersecurity firm SlowMist, the hack was possible due to the mismanagement of the access rights between two vital Poly Network's smart contracts. Although the hackers eventually returned the funds to Poly Network for fear that their own identity (such as IP address, email, and their account on centralized crypto exchanges) would be discovered and they could be prosecuted, the same thing could happen again without the return of funds, if hackers take more care of their own identity.

The good approach to secure smart contract development is to have a robust internal security review process to review smart contract code whenever there is any change in the code. Also, it is very important to hire at least two independent external smart contract audit firms to audit the smart contract before it is deployed on the mainnet. The cost of auditing and verifying smart contracts is high, and even after auditing, it is difficult to completely avoid security risks. Given the severe consequence of attacks, the funds spent on the audit is still worth its value.

Although there are many static code analysis and formal proof verification tools (both open source and commercial tools) available in the market, most of the effort to audit smart contract code is via manual code review by an experienced smart contract security auditor. Therefore, it is important for any metaverse project to hire a smart contract auditor in house for ongoing security reviews during code development, as well as to engage third-party auditors before the project goes live online.

Consensus Algorithm Security

Consensus algorithm is the root of the blockchain network. It enables the network nodes to agree on the state of the ledger. If the consensus algorithm is attacked, then the blockchain is not secure, and if there is monetary value stored on the chain, the economic loss can be devastating to the participants of the blockchain ecosystem. For example, in 2019, one consensus algorithm called "Sync Hotstuff" was found to have a critical security vulnerability with which an adversary, like so-called "force-locking attack," can conduct double spending or denial-of-service attack.

Another example is the attack on the Gasper algorithm authored by Vitalik Buterin and others. Gasper is an abstract proof-of-stake consensus layer that is implemented by the Beacon Chain protocol, the underlying protocol of the upcoming Ethereum 2.0 network. A key component of Gasper is a finality mechanism that ensures durability of transactions (safety) and the continuous operation (liveness) of the system even under attacks. It combines the finality gadget Casper FFG with the LMD GHOST fork choice rule and aims to achieve the safety and liveness. However, a 2020 paper authored by researchers from Stanford University has formally proved that an attack can be launched against Casper to impact its safety and liveness.

Node Security

Blockchain nodes are also called consensus nodes. They are distributed around the world to verify transaction signatures and use consensus algorithm to keep the ledger updated. Node is composed of the node software and the hardware (or microservice environment) that provides an execution environment for the node software to run. Node security includes the node software security and the hardening of the node execution environment.

For node security, one good example is EOS node software. EOS blockchain was a well-known blockchain developed by Block.One. A Chinese internet security research firm Qihoo 360 discovered a critical bug on EOS producer node that could be used by hackers to manage code on nodes remotely. The official blog post on Qihoo 360's website reads:

> This vulnerability could be leveraged to achieve remote code execution in the nodeos process (operating system), by uploading malicious contracts to the victim node and letting the node parse the malicious contract. In a real attack, the attacker may publish a malicious contract to the EOS main network.

Luckily, the vulnerability was fixed before EOS went live and there was no money lost. If this vulnerability had not been patched and EOS went live, billions of dollars' worth of funds would have been at risk.

Another good example of node security is the execution environment and firewall setup for the node to run. In a 2018 case, port 8545 was attacked, which is the default listening port for the Remote Procedure Call (RPC) interface of Ethereum clients, including Geth. As background, all Ethereum clients have a built-in RPC interface, which can provide third-party access via an API, thus possibly exposing sensitive information and operations.

By default, most Ethereum clients deactivate RPC, but users interested in enabling remote Ethereum blockchain access can activate the JSON-RPC interface. While authentication and Access Control Lists (ACLs) are supported, the interface can expose users' miner information and wallet details if connected to the internet. The hackers were able to drain about $20 million worth of ETH from the miners' address by using a

JSON-RPC call to miners' port 8545, because the miners misconfigured their loopback addresses and exposed the port to the internet.

In summary, for node security, it is important to conduct security testing for the node software to hunt for vulnerabilities before the node software can be used in the production environment (or mainnet). For the node execution environment, developers must make sure the operation system and microservice environment are locked down and hardened, the unnecessary services are shutdown, and the open ports are closed to allow only necessary ports to open and apply access control to the port. Center for Internet Security (CIS) benchmark can be a good reference for host environment hardening.

Data Encryption Security

Blockchain has some degree of data encryption to protect users' privacy. For example, the address in the blockchain system is generated by the user and has nothing to do with the user's identity information. There is no need for a centralized third party to participate in the creation and use of the address. Therefore, compared to traditional accounts (such as bank card numbers), blockchain addresses show better anonymity.

However, users may leak some sensitive information when utilizing the blockchain. For example, the broadcast of the blockchain transactions at the network layer may be used by hackers to infer the user's IP address, public key, or wallet address on the unrelated blockchain transactions. And Big Data analysis can be applied to find out the user's actual identity. Furthermore, since data on the public blockchain can be viewed by anyone, it is important not to put any sensitive information such as personal identifiable information, financial records, or medical records on the chain.

For cross-verification purposes and data integrity checking, it is a good approach to put the hash of the sensitive data on a chain since hash is one way to encrypt. It is easy to get a hash (also called digital digest) of the data, and it is mathematically impossible to derive the plain text data from the hash if the hash algorithm is strong enough. A variety of other cryptographic algorithms such as zero-knowledge proof, secure multipart computing, and homomorphic encryption have gained traction as new tools for data encryption. (Chapters 7 and 8 will cover data privacy and data security topics in detail.)

Five Challenges of Blockchain Adoption and Possible Solutions

The current blockchain technology are facing five main challenges: privacy, scalability, consensus algorithms, the authenticity of data on the chain, and interoperability.

Privacy Issues

From a privacy perspective, current blockchain projects are not mature enough. In the field of privacy protection, zero-knowledge proof, secure multi-party computing, homomorphic encryption, ring signature, BLS signature, Schnorr signature, Mibble Wimble, and other privacy algorithms are worth looking into. Research on state channels with access control and Trusted Computing Environment (TEE) has been very active, which also contributes to the development of privacy protection technology.

The challenge facing researchers is to meet the privacy requirements of the EU GDPR (General Data Protection Regulation), United States Privacy Laws like HIPPA (Health Insurance Portability and Accountability Act), and other regulations, as well as to meet the requirements of KYC/AML. There needs to be a balance between privacy protections and KYC/AML regulatory requirements.

Scalability

Can blockchain – a technology that started out as a niche project between enthusiasts – successfully scale to a global level? It's a big question. We can liken it to the early days of the internet, where the technology was grappling with a radically increasing user base and the challenges and slowdowns associated with that. From a scalability point of view, there are three-layer solutions. We use Ethereum scaling solutions as an example (see **Figure 2.4**).

Layer 2 Technology

The top layer (known as layer 2 technology) uses a different network running on the top of the main Ethereum network or layer 1. The Ethereum layer 2 solutions stay on the Ethereum network in the form of smart contracts. The layer 2 solutions don't need any modifications in the base level protocol for interacting with the main network. Ethereum layer 2 scaling solutions could serve different functions such as off-chain computation and scalability of payments.

The work of all layer 2 solutions focuses on one distinct element (i.e., moving the majority of the transactions off the

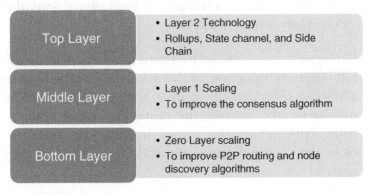

Figure 2.4   Ethereum Scaling at Three Layers

chain). As a result, layer 2 solutions could improve transaction processing speed while also reducing gas fees required for the transactions. Many Ethereum layer 2 solutions have been successful in gaining attention. Three main ideas of scaling on layer 2 are explored: rollups, state channel, and side chain.

Rollups. Rollups scaling solutions execute transactions outside of the Layer 1 blockchain and post the data from the transactions on it. Since the data is on the base layer, it allows Layer 1 to keep rollups secure. Rollups have two different security models:

- **Optimistic rollups.** These assume transactions to be valid by default. Thus, they only conduct computation to detect fraud if there's a challenge.
- **Zero-knowledge rollups.** These rollups run computations off-chain. Subsequently, they submit the validity proof to the base layer or mainchain.

Rollups, in the example of Polygon (formerly Matic), Arbitrum, and Optimism, help to increase transaction throughput and open participation, and they reduce gas fees for users.

State Channels. State channels allow two-way communication between participants of the blockchain to take place. In doing so, participants can reduce waiting time since there's no third-party – for instance, a miner at layer 1 chain – involved in the process. Here's how it works:

- Using smart contracts, the participants pre-agree to digitally sign off a portion of the tokens from the base layer.
- They can then directly interact with each other, eliminating the need to involve the miners at layer 1 chain.
- After conducting the entire transaction set, they can close the channel and commit the final state to the base layer blockchain.

Sidechains. A sidechain is a separate chain facilitating a large number of transactions. It has a consensus mechanism that's independent of the layer 1. The mechanism can be optimized to enhance scalability and processing speed. In this situation, the main chain must confirm transaction records, maintain security, and handle disputes. Sidechains differ from state channels in that they publicly record all transactions in the ledger. Also, if a sidechain experiences a security breach, it doesn't impact other sidechains or the layer 1 mainchain itself.

Layer 1 Scaling

The middle layer is known as layer 1 scaling. In this layer, the idea is to use sharding or segregated witness (Segwit), increase block size (i.e., increasing the amount of data contained in each block), reduce block confirmation time, use directed acyclic graph (DAG), or improve the consensus algorithm on the base chain to improve the performance and scalability of the base chain.

Zero Layer Scaling

The bottom layer is also known as zero layer scaling. In this layer, the main idea is to improve P2P routing and node discovery algorithms to obtain better scalability, or in the future, to use 5G technology to obtain better network bandwidth. Generally speaking, improvements in all three layers are necessary. All three layer technologies are under active research to improve the scalability of the blockchain.

Consensus Algorithms

From the perspective of consensus algorithms, the FLP impossibility theorem in the field of asynchronous communication has proved that in a distributed environment of completely asynchronous communication, if one node fails, the entire network cannot reach a consensus. Therefore, as consensus algorithm researchers try to bypass FLP impossibility, most consensus algorithms assume an honest majority of network nodes and partial (or complete) synchronization.

For example, the POW algorithm used in the Bitcoin Proof of Work algorithm assumes 51 percent honest nodes and the upper limit of response time (partial synchronization), and various types of POS algorithms also assume a majority of honest nodes and varying degrees of synchronization.

The key problem of the POS algorithm is the so-called *nothing at stake* attack. The nothing at stake relationship means that because the POS voting cost is almost zero, if there are multiple forks in the blockchain, each verifier will vote on all the forks to get rewards on all forks. In addition, POS also faces long-range attacks and other attacks. Using a game theoretic approach and introducing penalties for malicious POS nodes can somewhat mitigate POS vulnerabilities.

There is still much active academic and industry research on the POS front to make it secure. Examples of such research include the Ethereum 2.0 Proof of Stake algorithm and Cadano's Ouroboros Proof of Stake. Other consensus algorithms are still under active research. We see this as one of the major challenges for blockchain technology, because it is hard to reach security, scalability, and decentralization all at the same time. This is called the *blockchain trilemma*.

The blockchain trilemma involves three competing concepts. You can always achieve the three main attributes of scalability, security, and decentralization at the expense of others, but you cannot maximize all three properties at the same time. As such, the consensus algorithm design needs to find a balance among security, scalability, and decentralization based on the actual business use cases. For this reason, there will be no single blockchain to dominate all applications. There will be different chains with different consensus algorithms for different use cases.

Authenticity of Data on the Chain

Data authenticity means that data has not been corrupted after its creation and must represent a real-world scene. Data authenticity also means that a digital object is indeed what it claims to be or what it is claimed to be.

In order to make blockchain technology useful for real-world applications, there is a strong need to provide authentic data to the blockchain. Although blockchain has the immutability of data on the chain, it still faces the "first-mile" problem of mapping the attributes of real-world physical objects onto the chain. If there is no data authenticity, the smart contract on the blockchain can operate on fake and junk data, so the result of executing the smart contract may lead to asset loss or other serious consequences.

There are many approaches to obtain authenticated data for the blockchain, including the following:

1. The most used is the so-called "oracle" technology, which is mainly used to provide trusted data for DeFi applications (discussed in **Chapter** 4), such as the market price of bitcoin, the market price of stocks, and the exchange rate of the US dollar. The main drawback of this technique is that there is no effective trustless oracle scheme. The introduction of penalty mechanisms and digital identity mechanisms may lead to a better solution. Using decentralized identity as an oracle is a potential solution, in addition to a penalty mechanism for wrong oracle data.
2. For high-end sensors or servers, the trusted computing environment (TEE) is used to digitally sign the collected data. Because TEE physically protects the private key used for signing, the private key becomes invalid without the physical subject. This ensures that the data is not tampered with during transmission. Similar technologies include Secure Element, Physical Unclonable Function, and other techniques.
3. For assets with unique physical properties that can be measured, the data authenticity solution is simpler. Everledger, for example, uses the unique physical

properties of diamonds to record valuable diamond data on the blockchain, making natural diamond industry robust enough to cope with the growth of synthetic alternatives.

4. Using artificial intelligence, machine learning algorithms to identify fake data and increase the authenticity of chained data.
5. Monitoring IoT and other data collection assets with video surveillance and other security controls.
6. Implementing specialized chips (usually a security element chip with NFC function) into the physical objects at a random position to minimize the alternation, tempering, or counterfeit of physical objects.

Interoperability

Blockchain *interoperability* includes the ability to share, and invoke smart contracts from different blockchain networks without the need for an intermediary or central authority. Metaverse applications will use different blockchains of varying characteristics (governance rules, blockchain technology versions, consensus algorithms, permission controls, etc.), but separate blockchains do not work together, and there is currently no universal standard to enable different networks to communicate with each other.

The lack of interoperability can make mass adoption of blockchain in the Metaverse platform more difficult. The good news is that over the past few years we have seen an increasing number of interoperability projects striving to bridge the gap between different blockchains. Many of them aim to connect private networks to each other or to public blockchains. Examples of cross-chain projects, including Polkadot, Cosmos, and many others, have had varying degrees of success. (In Chapter 10 we will cover the latest development on interoperability, which will enable the open Metaverse to compete with Big Tech platforms.)

Why Blockchain Is Essential for Metaverse

This chapter discussed the features of blockchain technology and the convergence of blockchain and other digital technologies for metaverse applications. Now, we can appreciate why blockchain is a key enabling metaverse technology. As illustrated in **Figure 2.5**, blockchain technology can empower the Metaverse in eight major aspects.

1: Real-Time Global Payment Rail

Metaverse economies will need a global payment rail that can settle and clear the transaction in real time. The traditional payment system such as SWIFT, which still relies on batch processing of transaction settlement and clearing, will not meet metaverse payment requirements. Blockchain consensus algorithm enables the real-time settlement, since participating consensus nodes of the blockchain network use algorithms to settle financial transactions, without having to wait for a batch process to perform reconciliation and clearing tasks among different financial institutions, who have their own copy of financial records. (See detailed discussion in **Chapters 3 and 4**.)

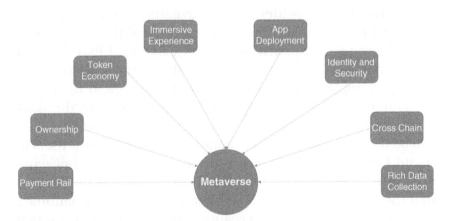

Figure 2.5   Blockchain Technology Empowers Metaverse

2: Ownership Verification for Digital Assets

For metaverse applications to create values in the system, the ownership of digital assets, which have no real-world presence or physical analogs, is essential. Ownership verification can be implemented, leveraging the immutability of blockchain's distributed ledger system and NFT (nonfungible tokens). Once the NFT is "minted," the transfer or sell of NFT is recorded on chain, and one can use on-chain transaction and owner's public key to validate the ownership of the NFT. The ownership of NFT is secured by the owner's private key. (See detailed explanation of NFT in Chapter 5.)

3: Crypto Tokens to Power the Creator Economy

How can we incentivize UGC (user generated content), participation, and governance of the Metaverse ecosystem? Crypto tokens and associated smart contracts have been used successfully to prompt the participation and governance of blockchain protocols.

In metaverse applications, token design will bring new variables and dimensions to encourage sustainable value creation and user participation. In general, the metaverse economic design needs to consider various dynamic variables such as number of users, token monetary policies, different kinds of token in the systems such as ERC20, ERC721, ERC1155, and the interactions of these tokens inside the Metaverse ecosystem.

For example, in a play-to-earn (P2E) game and metaverse integrated platform, the NFT can be used to reward the skills archived and time spent to play the game. The ERC20 token can be the basic currency of a metaverse system to allow users to spend ERC20 platform token to mint new NFT or other types of tokens. Careful economic design needs to be in place so that the platform token can arrive at a sustainable inflationary target to encourage value creation but discourage too much issuance or inflation. (See detailed blockchain gaming discussion in Chapter 6.)

Nevertheless, a pure deflationary platform token will make the token too expensive so as to discourage new users from participating in the Metaverse. So, a good token design will need a good monetary policy. The Metaverse's economic design team is similar to the Federal Reserve in the real world and the mission of the economic design team is to get a targeted and controlled level of inflation to encourage user participation and value creation. The token economic design also requires the revenue generated from existing users, not just solely from new users – to avoid Ponzi schemes.

4: Blockchain /AI to Create Immersive Experience

Blockchain can encourage, via incentive mechanisms written in smart contracts, the sharing of high-quality learning data and algorithms for AI (artificial intelligence) /ML (machine learning) among metaverse participants, and the AI/ML can create rich and immersive experiences. AI can create human-like voices and unique contents. The data can be automatically converted into games, videos, news, advertisements, and lecture materials by using some sample learning data shared via incentive mechanisms. It is possible for AI to create extensive content that imitates human behavior by using the vast data in the Metaverse world. (This relates to the talent bottleneck of the creator economy discussed in Chapter 10.)

5: Decentralized Cloud for App Deployment

Metaverse applications can be deployed in the cloud, IPFS, and also on blockchain. The UI (user-interface) contents, like video and audio contents, can be stored in IPFS or highly efficient AWS or Microsoft Azure cloud. The smart contract can be deployed on Ethereum mainnet or layer 2 roll-up blockchain or other Ethereum alternative blockchains such as Binance Smart Chain, Polygon, or Avalanche. In the long run, the blockchain-powered cloud computing environment will provide more secure and cheap alternatives to AWS and "Big Tech clouds." (See related discussion in Chapter 2.)

6: Decentralized Identity and Cybersecurity

The current internet is based on TCP/IP, which does not include the specification and implementation of identity security and cybersecurity. Metaverse can leverage blockchain for data integrative, smart contract execution security guarantee, and decentralized identity for data ownership. The common cybersecurity practices such as zero trust, API security, and access management can further leverage blockchain to build foundational support for metaverse applications. (We will discuss metaverse security in detail in Chapter 8.)

7: Cross-Chain Computing Turns "Multiverse" into Metaverse

For the Metaverse to flourish, it must be an open and interoperable system. Cross-chain efforts such as Cosmos and Polkadot can be further enhanced to support interoperable metaverse applications. To ensure data privacy for cross-chain metaverse transactions, privacy preserving technologies (we will discuss more about this in Chapter 7) can also be leveraged. (Interoperability will be further discussed in Chapter 10.)

8: Enable New Data Economy in the Metaverse

Increasing interest in the Metaverse has coincided with an explosion in the development and use of new digital methods of exchanging virtual 3D assets such as virtual land, Pokémon cards, and gaming weapons. The key to this paradigm is the blockchain technology that enables users to verify the authenticity of the digital asset being sold. And this will eventually start to influence how AR apps are developed and turned into profitable ventures. For example, Cappasity is a decentralized AR/VR ecosystem for 3D content exchange, using blockchain to allow 3D content creators to produce, rent, and sell AR/VR content through the Cappasity market-place. Each asset is assigned a unique identification code to prevent copyright infringement.

In the 3D interactive context, the Metaverse is a huge new source of user and behavioral data. In addition to browsing and transaction data, metaverse players use new hardware such as AR/VR headsets and IoT devices to collect more data than ever before. Some of that data will be very valuable and have a privacy impact if data governance is not in place. Blockchain technology can be used to define the ownership of data, enable price discovery of data, as well as facilitate data exchange in a privacy preserving fashion. Real-time data analytics is the new paradigm for organizations of all kinds as we slowly build up the Metaverse. (See related discussion in Chapters 6, 7, and 8.)

In summary, blockchain is the backbone of Web3. In Part II of this book, we will discuss the cutting-edge blockchain break-throughs that are setting the transaction, privacy, and security foundation for this digital economy.

PART II

Blockchain Breakthroughs Set the Transaction, Privacy, and Security Foundation for the Digital Economy

Just like blockchain is (way) more than Bitcoin, Web3 is expanding beyond its financial origins to become the new internet based on ownership and decentralization. The interplay among crypto, DeFi, NFT, gaming, and social work are driving more tech innovation and user cases in the blockchain-based creator economy.

Cryptocurrencies and Tokenomics

- Virtual Lands, Virtual Currencies
- Why Metaverse Needs Blockchain-Based Transactions
- Bitcoin: The Beginning of Cryptocurrency and Trust
- Ethereum: Smart Contract Execution Platform
- Bitcoin versus Web3 (Pure Decentralization vs. Modified Decentralization)
- Visa and Mastercard: Tokenomics Going Mainstream
- Metaverse Outlook: Crypto Beyond Currencies

Virtual Lands, Virtual Currencies

Virtual land sales in the metaverse have skyrocketed since Facebook's shift to Meta in 2021. According to a December 2021 *Business Insider* article, investors are describing virtual land sales now as "buying on 5th Avenue back in the 1800s," and land parcels in various metaverse sites are being snatched up fast.

Some of the priciest deals include that of Republic Realm for a $4.3 million purchase of land from video game publisher Atari SA, in the Sandbox metaverse in November 2021. Similarly, in November 2021, according to the *Strait Times,* a parcel of land in the online Decentraland metaverse was sold for a record $2.4 million to the Canadian cryptocurrency company Tokens.com. At the time of transaction, Decentraland has said that to date, it was the most expensive purchase of a parcel of virtual real estate on the platform. Metaverse Group, a subsidiary of Tokens.com, carried out the purchase using Decentraland's own cryptocurrency, MANA, at 618,000 MANA (equivalent to $2,428,740 at time of purchase).

The virtual land sale phenomenon is not limited to the aforementioned platforms. According to cryptocurrency data site Dapp.com, the four largest metaverse websites – the Sandbox, Decentraland, Cryptovoxels, and Somnium Space – sold virtual real estate valued at more than $100 million in the first week of December 2021.

Investors have cited a myriad of reasons for their purchases, including building immersive experiences for their customers (similar to building/renting stores in a physical central business district), or speculating on the value of what could become the future meta-capitals of the world. It could be argued that with the advent of an industry heavyweight such as Facebook becoming Meta, speculators understood the move as an industry giant affirming that the metaverse is its main growth direction, taking its ~2.9 billion monthly active users with it (as of 2021 Q2).

So, we start Part II with the most obvious use cases for blockchains in the Metaverse – money. Blockchains are the foundation of cryptocurrencies like bitcoin, ether (the token of the Ethereum blockchain), as well as the cryptocurrency mana that metaverse enthusiasts buy plots of virtual real estate within the Decentraland online reality (the Decentraland base will be covered in Chapter 6 in connection with blockchain gaming). As well as land, metaverses make it possible to buy digital versions of just about anything we can buy in the real world.

However, until crypto and blockchains existed, online social spaces like Second Life were not considered a *real* metaverse because all the value it generated was locked inside its centralized servers. For the users, they could not take their belongings from Second Life and carry them as they traveled to different destinations in the metaverse. Thanks to the blockchain's ability to work as a universal language for proving a digital object's provenance, crypto makes this possible. Without further ado, let us first explain why blockchain technology is a critical piece of the metaverse puzzle.

Why Metaverse Needs Crypto and Blockchain-Based Transactions

What advantages does a monetary transaction system on blockchain offer? Think of blockchain as a database shared across many participants, each with a computer. At any moment, each member of the blockchain holds an identical copy of the blockchain database, giving all participants access to the same information. As such, all blockchains share three attractive characteristics:

1. **A cryptographically secure database (also known as distributed ledger).** That means that when data is read or written from the database, you need two cryptographic keys to do it: a public key (basically the address), which is essentially the address and the database where information is stored, and a private key, which is your personal key, truly the security provider that prevents other people from updating the information unless they have that correct key. Users cannot update the blockchain unless they have the correct keys. Blockchain's cryptographic keys provide leading-edge security that goes far beyond that found in a standard distributed ledger. The technology also eliminates the possibility that a single point of failure will emerge since the blockchain database is distributed and decentralized. If one node fails, the information will still be available elsewhere.
2. **A digital network that enables the sharing of a digital log of transactions.** Transactional information is available in real-time through the blockchain network. The most famous public network is the bitcoin blockchain, which will be discussed in detail in this chapter. A blockchain

network can either be a "public network" or "private network" – anyone can join or leave a public network without express permission, whereas admission into private networks is by invitation only. In a public chain, you can join and leave again and again, and no one really knows who's joining and leaving.

For example, IDC unveiled a long list of industries where blockchain could foster trust including government, healthcare, logistics, shipping, and of course, finance, where it originated. The analysts predicted that by 2022, healthcare blockchain digital identity standards will come from the US Department of Health and Human Services, enabling universal medical data interoperability, comprehensive customer/patient data, patient "health scores," and AI. They also saw blockchain-enabled electronic voting eventually emerging; 8 percent of jurisdictions worldwide will test systems by 2023.

In the same time frame, IDC analysts predict that 65 percent of transcontinental shipping will be legislated to use blockchain that encompasses crew health information, bunker fuel sourcing, and goods origination data. By the next year, 15 percent of supply chain transactions will use blockchain for the provenance of ethical and sustainable practices to increase digital trust. By 2025, these analysts said that 10 percent of financial institutions will use blockchain technology for know-your-customer (KYC) compliance to create a transparent, auditable record of entities.

Why is this exciting for the development of the internet – into Web3.0? After three decades, most internet applications are links between just two points at a time, even at high connection speeds a choke point on potential uses. Blockchain's selling point is to eventually allow numerous parties in a transaction to interact simultaneously and securely transfer assets over the net. To replace email chains and paperwork in a house sale, for example, the buyer, seller, and brokers would tap into the same blockchain system as lawyers, mortgage bankers, and title examiners.

3. **The audit trail created by the blockchain networks.** The distributed ledger technology records all transactions between multiple parties on one theoretically immutable chain. The database can only be updated when two things happen. First, a user must provide the correct public and private keys. Second, a majority of participants in the network must verify those credentials. This reduces the risk that a malicious user will gain illicit access to the network and make unauthorized updates.

Since everyone on the chain can immediately see all the data, including all transactions, this reduces the risk for fraud. For example, in the context of real estate transactions, users can go back through the blocks of information and easily see the information previously recorded in the database, such as the previous owner of a piece of property. Also, it is easier for companies to prove compliance with regulations and head off expensive audits.

Smart Chopsticks and Blockchain Chickens

Because of the widespread food scandals, Chinese consumers' confidence in food, especially when it is domestically produced, is low. In 2008, milk powder tainted with melamine, a toxic industrial compound, made 300,000 babies ill and killed 6 as a result of consumption of the tainted milk powder. Since then, most Chinese parents have turned away from locally produced brands, and the supermarkets in Australia, Japan, and Hong Kong often saw Chinese tourists buying up the baby formula on their shelves.

More stomach-churning food scandals followed the deadly tainted milk case. Chinese consumers have also encountered watermelons that exploded from the misuse of a growth accelerator chemical, lamb made of rat meat, pork soaked in a detergent additive, and cooking oil recycled from waste oil collected from restaurant fryers, grease traps, or even sewer drains (known as the "gutter oil"). Sometimes it is just hard to figure out how far the food is away from "fresh." In 2015, Chinese authorities seized 3 billion RMB (close to

$500 million) worth of frozen beef, pork, and chickens that dated as far back as the 1970s. "A bottle of 1982 Lafite plus a piece of 1970s steak and a pair of 1980s chicken wings," wrote one Chinese user on the Sina Weibo microblog (the Chinese equivalent of Twitter), "Bon appétit!"

On April Fools' Day 2014, China's search engine giant Baidu offered a video clip on smart chopsticks that could determine whether a dish contained gutter oil. According to Baidu, when the video was made, it had no serious intention to pursue it as a product. However, because the fake advertisement generated so much buzz on social media, Baidu decided it could be a timely innovation.

At the company's annual technology conference in September 2014, Baidu's Chief Executive Robin Li unveiled the "smart chopstick" prototype that was called Baidu Kuaisou. They were equipped with sensors to collect data on pH levels, peroxide value, and temperature, and they could be connected to a smartphone app to provide users with analyzed readings on the oil being tested. But some food expert immediately warned that gutter oil producers could outsmart the smart chopsticks. Because the sensors only take a small number of variables for its analysis, the gutter oil producers could, according to the experts, easily add relevant chemicals to give its oil products a false safe reading.

Now the cutting-edge technology for food security is blockchain. It is used to collect data about the origin, safety, and authenticity of food and provide real-time traceability throughout the supply chain. This has traditionally been challenging due to complex and fragmented data sharing systems that are often paper-based and can be error-prone. The blockchain solution would provide consumers and regulators far more information about the food on the shelves: its source and region, its shipping process, and its inspection and certification, among other useful information as well.

For example, JD.com, a major e-commerce competitor of Alibaba, piloted a blockchain application for consumers in select Chinese cities to track meat from Chinese beef producer Kerchin based in Inner Mongolia. Working with the beef producer, JD allowed consumers to access detailed information, such as the cow's breed, when it was slaughtered, and what bacteria testing it went through. Furthermore, it worked with Australian exporter InterAgri to use blockchain to track the production and delivery of Black Angus beef from import.

Tech firms, always in a growth mindset, thought they could go even further with blockchain. In 2019, GoGo Chicken, a poultry monitoring technology based on blockchain, was developed by ZhongAn Technology, a subsidiary of the Chinese online insurer ZhongAn Online (Alibaba is a shareholder), to chronicle a chicken's life story to prove it is organic (or not). According to the company, each chicken would wear a tracking device on its foot, which automatically uploads its real-time movements through the supply chain to the blockchain database. Sensors monitor temperature, humidity, and other aspects of the chicken's environment, while algorithms evaluate the bird's health using video analysis. ZhongAn plans to roll it out to hundreds of Chinese farms by 2020 and is confident that eco-conscious consumers will be happy to pay a premium to ensure that the chickens they buy are truly cage-free.

In short, blockchain, a technology enabling the decentralized and secure storage and transfer of information, could become a powerful tool for tracking and transactions that can minimize friction, reduce corruption, increase trust, and empower users. Blockchain is the technology that allows cryptocurrencies to change hands online without assistance from banks or other intermediaries.

During the last decade of mobile internet and smartphone-led "mobile revolution," internet users have already become used to buying consumer goods anytime and anywhere. Now the metaverse is pushing that trend further – from consumer goods to new digital assets, and the blockchain-based cryptocurrencies are the access to the trillions of future transactions. In the following section, we will start the cryptocurrencies discussion with Bitcoin, the origin of all cryptocurrencies.

Bitcoin: The Beginning of Cryptocurrency and Trust

In 2011, a bitcoin was worth $1. At the end of 2021, it was trading at about $50,000, and the value of all cryptocurrencies, of which bitcoin is one among many, is some $2.3 trillion. The total value of the cryptocurrency market, including bitcoin, ether (which will be covered in detail in the following section), and various smaller tokens, was $3 trillion during November 2021's all-time high. At the time, bitcoin's price approached $69,000 and ether came close to $4,900 (see **Figure 3.1**).

In short, crypto officially entered the mainstream as bitcoin and ether hit new all-time highs in 2021. This rise has led many to envision a radically different future for finance and to question long-held beliefs about value. Blockchain – where bitcoin is created (or "mined") upon – has become a new way of looking at value and a new way of creating a transaction between parties where you don't need a third-party intermediary and can track things and really have trust.

A short history of bitcoin might provide a good overview. Most popular accounts for the beginning of bitcoin put that date on August 18, 2008, when the domain name bitcoin.org

Bitcoin Value's Growth Journey (2010 – 2022)

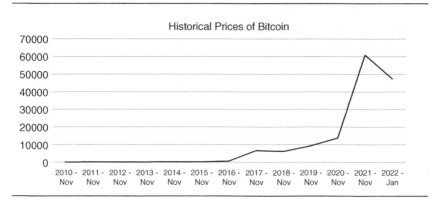

Figure 3.1   Historic Price of Bitcoin
Sources: in2013dollars

was first registered. On October 31, 2008, a developer under the pseudonym Satoshi Nakamoto published the bitcoin White Paper via metzdowd.com's cryptography mailing list, titled "Bitcoin: A Peer-to-Peer Electronic Cash System." Several major sites began to accept bitcoin as a payment currency, such as WikiLeaks in 2011, and WordPress in 2012. In February 2013, Coinbase reported that it was selling $1 million worth of bitcoins in a single month at over $22/BTC.

During the past decade, governments worldwide shifted in opinion towards bitcoin as bitcoin's popularity grew. In 2016, the Cabinet of Japan approved a set of bills to help banks expand their business and recognize virtual currencies – such as bitcoin – as having functions similar to real money. Russia similarly moved to legalize the use of cryptocurrencies in 2017 as reported by *Business Insider*, creating the legal framework for trading in currencies such as Bitcoin and Ether. (See **Figure 3.2**. The tension between decentralized cryptocurrencies and government regulations will be covered in **Part III** of this book.)

Satoshi Nakamoto's paper marketed the idea of bitcoin as a decentralized currency that is devoid of presence of central

Key Milestone Events of Bitcoin

2008–2009:	2010–12:	2013–14:	2015–16:	2017–18:	2019–20:	2021:
Satoshi Nakamoto publishes Bitcoin white paper; Genesis block created and mining commences; Bitcoin vo.1, vo.2 released	Bitcoin vo.3, released; total market cap tops $1M; Bitcoin reaches and fluctuates at $1/btc; bitcoin exchanges suffer hacks; Wordpress receives bitcoin	Bitcoin worth $100/btc; Mt. Gox exchange bankrupts; Bitcoin Core version 0.9.0 is released	100K+ retailers worldwide accept bitcoin; Japan passes bill recognizing virtual currencies as similar to real money; Bitcoin halving event	Japan officially recognizes Bitcoin as legal tender; China closes all domestic cryptocurrency exchanges; Bitcoin worth $20K/btc; Bitcoin Core 0.17.0 becomes available	NYSE launches market for Bitcoin futures; China supports blockchain; COVID-19 drives bitcoin to fall by 50%; Ethereum catches up in value settled	Bitcoin worth $50K/btc; Morgan Stanley offers Bitcoin access; Coinbase goes public; China cryptocurrency crackdown; EL Salvador makes bitcoin legal tender

Figure 3.2   Key Milestone Events of Bitcoin
Source: Medium, TradingView, HowMuch.net

banks, and as "a purely peer-to-peer version of electronic cash...
on a network that requires minimal structure." Loughborough
University's Honorary Fellow Dave Elder-Vass described Nakamoto's vision of bitcoin transactions as: "anonymous, safe, more
or less, from the prying eyes of the state." Nakamoto released
the very first Genesis Block (also known as Block 0) of bitcoins
on January 3, 2009, the original block containing the first 50
bitcoins, via the SourceForge platform. Block 1 was mined on
January 9, 2009, six days later from Block 0.

As illustrated by Bitcoin, blockchain drastically reduces
the time and resources to verify transactions. A blockchain is a
public database that is updated and shared across many computers in a network. *Block* refers to data and state being stored
in consecutive groups known as "blocks." *Chain* refers to the
fact that each block cryptographically references its parent. In
other words, blocks get chained together. Every computer in
the network must agree on each new block and the chain as
a whole. These computers are known as *nodes*. Nodes ensure
everyone interacting with the blockchain has the same data.

To accomplish this distributed agreement, blockchains
need a consensus mechanism. The data in a block cannot

change without changing all subsequent blocks, which would require the consensus of the entire network. All the core elements of blockchain are designed to make the protocol impossible to fake or replicate. Each block's timestamp marks the date of any previous transaction, and the cryptographic hash in each block maps to the previous block, so that no single block can be changed without disrupting every other block (see **Figure 3.3**).

Blockchain: What it is, and what it can be used for

Question	Reality	Applications and Use Cases
1. What is blockchain? Is blockchain bitcoin?	Blockchain is not bitcoin, and blockchain technology can be used and configured for many applications beyond bitcoin.	**Record Keeping – Storage of Static Information** 1. Static Registry (e.g., patents) 2. Identity (e.g., identity records) 3. Smart Contracts (e.g., insurance-claim payout)
2. Is blockchain better than traditional databases?	Blockchain is not necessarily better than traditional databases, but it is valuable in low-trust environments where participants cannot trade directly.	
3. Is blockchain tamper-proof or 100% secure?	Blockchain could be tampered with if >50% of the network-computing *power* is controlled and all previous transactions are rewritten – which is largely impassable.	**Transactions – Registry of Tradeable Information** 1. Dynamic Registry (e.g., drug supply chain) 2. Payments Infrastructure (e.g., cross-border peer-to-peer payment) 3. Other (e.g., initial coin offering)
4. Is blockchain a "truth machine"?	Blockchain can verify all transactions and data entirely contained on and native to blockchain, but it cannot assess whether an external input is accurate or "truthful."	

Figure 3.3   Blockchain Myths vs. Reality
Source: Adapted from McKinsey & Company, "Blockchain explained: What it is and isn't, and why it matters"

With bitcoin, the first participant, or "miner," to validate a transaction and add a new block of data to the digital ledger will receive a certain number of tokens as a reward. Under this model, which is referred to as a proof-of-work (PoW) system, miners have an incentive to act quickly. But validating a transaction does not simply involve verifying that bitcoin has been transferred from one account to another. Instead, a miner must answer a cryptographic question by correctly identifying an alphanumeric series associated with the transaction.

This activity requires a lot of trial and error, making the hash rate – the compute speed at which an operation is completed – extremely important with bitcoin. According to the cryptocurrency exchange Gemini, the PoW system has historically provided better security for users due to its decentralization and its requirement for miners to dedicate colossal amounts of computing power to validate transactions. It would require a single malicious actor with astronomical computing power and expense to even attempt controlling most of the computing power of the network. (In **Chapter 8**, the security consideration of bitcoin and other crypto assets will be discussed in detail.)

POW vs. POS

While the PoW system consensus mechanism has its merits, most notably perhaps its security feature, this very feature has become the reason for why market players have designed and are choosing its successor, the PoS system (proof-of-stake).

PoW means that anyone who wants to add new blocks to the chain must solve a difficult puzzle that requires a lot of computing power. Solving the puzzle "proves" that you have done the "work" by using computational resources, and this effort is known as "mining." (Hence the players are referred to as "miners.") Mining is typically brute force trial and error, but successfully adding a block is rewarded in bitcoins. In other words, when you send bitcoins to someone, the transaction must be mined and included in a new block. The updated state is then shared with the entire network.

However, the flip side of the PoW system's computing power requirement is that mining is so energy intensive that it has become environmentally damaging and economically exhaustive. Digiconomist's bitcoin Energy Consumption Index reports that the bitcoin cryptocurrency individually has a carbon footprint comparable to that of the country of New Zealand, producing 36.95 megatons of carbon dioxide annually.

Currently, Ethereum is considering upgrading to Ethereum 2.0 with replacement of its PoW system with one based on PoS. In a PoS system, participants are rewarded based on the number of coins they have in their digital wallets and the length of time they have had these stakes. Compared to the "miner" PoW consensus mechanism that came before, the PoS system adopts a "staking" consensus mechanism model. The participant that rates highest on these factors is chosen to validate a transaction and receive a reward. Many other large cryptocurrency networks, including Cardano, Dash, and EOS, are also investigating PoS algorithms.

The PoS systems have several advantages, among which are enhanced security, accessibility, and sustainability features. First, they help cryptocurrency networks build a trusted network of loyal participants – and this may make security breaches less common. Second, they level the playing field for cryptocurrency miners, since those with the greatest compute power will not necessarily be the winners. Players also appreciate that PoS systems are more energy efficient and allow faster transactions, lessening the energy burden exhaustive mining has on the environment.

The PoS system, however, is not without its drawbacks. Since PoS depends on a system of validators who stake a portion of transactions, validators involved could potentially go rogue and validate erroneous or conflicting transactions, at minimal cost to the validators themselves. As PoS does not have a requirement for massive amounts of energy, validators with less energy at their disposal (and more financial resources staked) could become dangerous to the whole system of transacting players. A shift to PoS systems could have major implications for semiconductor companies that serve cryptocurrency players, since it would shift chip demand in new directions. As to which out of the time-tested PoW or burgeoning PoS system will emerge as the majority mechanism adopted in the future, one still awaits how the market players will adapt and react.

As a result, the blockchain technology is most important as a way of verifying any transaction. The practical impact of this for a business lies in the time and money a streamlined verification process saves. This, however, has an underlying major ESG issue.

For example, in March 2021, Tesla CEO Elon Musk had announced on Twitter that the carmaker would accept the most popular and largest cryptocurrency, bitcoin, as a mode of payment to purchase electric vehicles. However, the electric-car maker halted car purchases with bitcoin in mid-May due to concerns over how cryptocurrency mining, which requires banks of powerful computers, contributes to climate change. After 49 days of accepting the digital currency, Tesla reversed course on May 12, 2021, saying they would no longer take bitcoin due to concerns that "mining" the cryptocurrency was contributing to the consumption of fossil fuels and climate change.

Musk said that the company would resume bitcoin transactions once it confirms there is reasonable clean energy usage by miners. "When there's confirmation of reasonable (~50 percent) clean energy usage by miners with a positive future trend, Tesla will resume allowing bitcoin transactions," Musk wrote in a tweet. During a July 2021 Bitcoin conference, Musk suggested Tesla could possibly help bitcoin miners switch to renewable energy in the future and also stated that if bitcoin mining reaches and trends above 50 percent renewable energy usage, that "Tesla would resume accepting bitcoin."

In addition to using renewable energy, another approach to reduce carbon emission is to use different algorithms to verify transactions on blockchains. Behind the scenes, more subtle changes are occurring in the cryptocurrency market as players try to minimize the importance of compute power by developing new algorithms, such as POS. (See **Box: POW vs. POS**.)

Ethereum: Smart Contract Execution Platform

After bitcoin, many of the additional cryptocurrencies – also called "altcoins" – were created to address certain gaps or inefficiencies with bitcoin, and they are available through various networks. Popular altcoins include Dash, Litecoin, and XRP (offered through Ripple). Of all the alternative cryptocurrency networks, Ethereum is most popular (see **Figure 3.4**). It is an open-source platform that allows users to build and launch decentralized applications, including cryptocurrencies or digital ledgers. Users must spend a specific digital currency, ether, to run applications on Ethereum. Ether can also serve as an alternative to regular money, but its primary purpose is to facilitate Ethereum operations.

As shown by the bitcoin section, blockchain is a distributed ledger, or database, shared across a public or private computing network. Each computer node in the network holds a copy of the ledger, so there is no single point of failure. Every piece of information is mathematically encrypted and added as a new "block" to the chain of historical records. Various consensus protocols are used to validate a new block with other

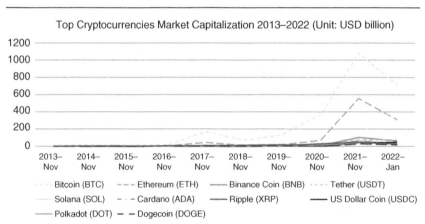

Figure 3.4 Market Cap of Major Cryptocurrencies (2013–2022)
Source: Forbes (2021, Nov 30), CoinMarketCap (2014–2022, Nov 30 for 2014–2021, Jan 30 for 2022), Statista (2013)
Notes: Polkadot was launched in May 2020, Solann Mar 2020, USDC May 2018, Binance Jul 2017, Cardano Sept 2017, Ethereum Jul 2015, Tether Jul 2014, Dogecoin Dec 2013. Data is not available prior to their founding date

participants before it can be added to the chain. This prevents fraud or double spending without requiring a central authority.

Represented by Ethereum, the blockchain ledger can also be programmed with "smart contracts," a set of conditions recorded on the blockchain, so that transactions automatically trigger when the conditions are met. For example, smart contracts could be used to automate insurance-claim payouts. It allows financial and other assets to be exchanged using computer code with no attorney or escrow agent involved. (As will be discussed in **Chapter 8**, smart contracts themselves are a new type of programming, and it takes careful diligence and review to ensure secure, gas-efficient smart contracts that reduce the risk of vulnerabilities, which we cover in our section on smart contract security.)

In 2014, Ethereum came in with a new proposition for building decentralized applications. There would be a single blockchain where people would be able to deploy any kind of program. Ethereum achieved this by turning the *Application* layer into a virtual machine called the *Ethereum Virtual Machine (EVM)*. This virtual machine was able to process programs called *smart contracts* that any developer could deploy to the Ethereum blockchain in a permissionless fashion. This new approach allowed thousands of developers to start building decentralized applications (dApps).

A smart contract is simply a program that runs on the Ethereum blockchain. It's a collection of code (its functions) and data (its state) that resides at a specific address on the Ethereum blockchain. At a very basic level, a smart contract can be viewed as a sort of vending machine: a script that, when called with certain parameters, performs some actions or computation if certain conditions are satisfied. For example, a simple vendor smart contract could create and assign ownership of a digital asset if the caller sends ether (ETH) to a specific recipient.

Smart contracts are type of Ethereum account, where accounts are defined as entities in the network that can hold a balance and send transactions. This means they have a monetary value balance, and they can send transactions over the network. However, they are not controlled by a user; instead they are deployed to the network and run as programmed. User accounts can then interact with a smart contract by submitting transactions that execute a function defined on the smart contract. Smart contracts can define rules, like a regular contract, and automatically enforce them via the code. Smart contracts cannot be deleted by default, and interactions with them are irreversible.

Ethereum currently uses a PoW consensus mechanism. EVM is essentially a single, canonical computer whose state everyone on the Ethereum network agrees on. Everyone who participates in the Ethereum network (every Ethereum node) keeps a copy of the state of this computer. Additionally, any participant can broadcast a request for this computer to perform arbitrary computation. Whenever such a request is broadcast, other participants on the network verify, validate, and carry out ("execute") the computation. This execution causes a state change in the EVM, which is committed and propagated throughout the entire network.

Requests for computation are called transaction requests; the record of all transactions and the EVM's present state gets stored on the blockchain, which in turn is stored and agreed on by all nodes. Any participant who broadcasts a transaction request must also offer some amount of ether to the network as a bounty. This bounty will be awarded to whoever eventually does the work of verifying the transaction, executing it, committing it to the blockchain, and broadcasting it to the network. The amount of ether paid corresponds to the time required to do the computation.

The ether cryptocurrency supports a pricing mechanism for Ethereum's computing power. When users want to make a transaction, they must pay ether to have their transaction recognized on the blockchain. These usage costs are known as gas fees, and the gas fee depends on the amount of computing power required to execute the transaction and the network-wide demand for computing power at the time. Because of the boom of crypto transactions in 2021, the Ethereum network was pushed to its full capacity, and the gas fees on Ethereum skyrocketed. As Web3 evolves, the demand for crypto infrastructure also increases.

Bitcoin versus Web3 (Pure Decentralization vs. Modified Decentralization)

As illustrated, blockchain tokenization and smart contract techniques can act as a form of advanced software that ensures that what has been agreed is executed upon, enabling the recognition of data rights, risks, and rewards. In the creation of economic value models, blockchain may act as an enabler in market-based mechanisms to attach value and permissions to the data.

But which crypto token will be *the currency* of the Metaverse? Is this about a war between Bitcoin and Ethereum? No doubt, bitcoin is the most recognized, first-ever digital asset, but there are hundreds and even thousands of other digital assets in the ecosystem. But there is also concern regarding bitcoin that as the first digital asset, it may be vulnerable to innovative destruction from competitors like Ethereum (such as the story of MySpace and Facebook).

In other words, is there going to be one clear winner that is crowned as the main metaverse currency? Or is it likely that there will always be a coexistence of multiple cryptocurrencies, each serving slightly, or entirely, different functions?

To address these questions, let's revisit the "Web3" moniker first. Web3, through blockchain technology, is hoped to enable users/consumers to own and thereby control their data and, by granting permission to use their data, own the platform and become "shareholders." The "shares" in Web3 are cryptocurrencies or tokens and represent ownership of the networks/the blockchains. By contrast, the current, main internet (Web2.0) is decided by a small number of tech giants, like Google, Facebook, Amazon, Twitter, and Apple.

Since cryptocurrency began, with the revolutionary possibility of delivering a decentralized platform for recording and transferring value, countless ambitious personalities have been seeking to decentralize and democratize the internet. The vision for Web3 is that, unlike the internet, through a specific cryptocurrency, which is the native "currency" to a particular crypto platform, one can actually invest in the platform itself (which is impossible in the case of the Web 2 internet) that will enable and support all of the value recording and transfer processing on the specific platform.

For the "purists" of Web3, their vision for the future internet is *totally decentralized*: (i) never allowing any single party or small group of parties to control or exert overwhelming influence over the platform ("trustlessness"); (ii) have extreme data security (the Bitcoin network cannot be hacked); (iii) be censorship resistant (anyone can use Bitcoin and nobody can stop any transaction by technical means); and (iv) ensure extreme privacy (although not necessarily secrecy).

Now we go back to the beginning of the blockchain story. The first blockchain was Bitcoin, a peer-to-peer digital currency created in 2008 that used PoW, a novel consensus mechanism. It was the first decentralized application on a blockchain, and at the same time, it is a *truly* decentralized platform, and the elements that have contributed to Bitcoin's success include:

- Anonymity of Bitcoin's creator (maybe the hardest for any other token to match)
- Fair launch (no "premine" or token allocation for early or any investors)
- Finite number of Bitcoin to ever be issued ("scarcity")
- PoW support, which is a more equitable distribution method than PoS, which is more widely employed and less energy intensive
- Relatively dispersed investor base

However, Bitcoin's "decentralized-first" nature cuts both ways. On the one hand, Bitcoin is fundamentally different from any other digital asset. No other digital asset is likely to improve upon bitcoin as a monetary good because bitcoin is the most (relative to other digital assets) secure, decentralized, sound digital money. In other words, any "improvement" will necessarily face tradeoffs.

On the other hand, Bitcoin is a rigid ecosystem that does not expand easily. That became obvious years ago, when people started to realize the potential of decentralized applications and the desire to build new ones emerged in the community. At the time, there were two options to develop decentralized applications: either fork the bitcoin codebase or build on top of it. However, the bitcoin codebase was very monolithic; all three layers—networking, consensus and application — were mixed together. Additionally, the Bitcoin scripting language was limited and not user-friendly. There was a need for better tools.

Hence the emergence of "smart contracts." Any developer can create a smart contract and make it public to the network, using the blockchain as its data layer, for a fee paid to the network. Any user can then call the smart contract to execute its code, again for a fee paid to the network. Thus, with smart contracts, developers can build and deploy arbitrarily complex user-facing apps and services such as marketplaces and games, and more. In practice, participants do not write new code every time they want to request a computation on the EVM. Rather, application developers upload programs (reusable snippets of code) into EVM storage, and users make requests to execute these code snippets with varying parameters.

Because ETH and smart contracts have empowered the spread of the crypto world, some part of the Web3 community argues for *smart-contract blockchain-based platforms* in the Metaverse, like, Ethereum, Solana, Cardano, Polkadot, and

Avalanche, which support many new and diverse cryptocurrency functions like:

- DeFi (decentralized finance)
- NFTs (nonfungible tokens)
- play-to-earn (P2E) gaming, using blockchain technology

These platforms allow for transparent, irreversible, and open-access transactions, each of which is an aspect and goal of decentralization. (The following chapters will discuss these "new" crypto functions in detail.) However, each of these platforms has founders, VC investors, and capitalist motivations for the success of a particular platform and its native cryptocurrency. In contrast, Bitcoin is decentralization maximized. As mentioned earlier, any "improvement" on Bitcoin will necessarily face tradeoffs.

In summary, as of now, bitcoin is the most pure, decentralized cryptocurrency, with the widest holdings and the greatest market cap. However, other cryptocurrencies (the "altcoins" or "alternative coins") have been gaining on bitcoin, in value and in the number of adopters. Because technological updates and changes cannot be deployed easily or quickly in connection with totally decentralized platforms, it makes sense that altcoins under the "control" of an active creator or group of developers can be expanded, aggressively marketed, and widely "sold."

As such, there is not necessarily mutual exclusivity between the success of the Bitcoin network and all other digital asset networks. Rather, the rest of the digital asset ecosystem can fulfill different needs or solve other problems that bitcoin simply does not. For the foreseeable future, we may see bitcoin and smart contract tokens coexist and co-evolve as Web3 evolves. At a point, the positive aspects of centralized projects will outweigh the negative aspects and the crypto ecosystem will migrate toward a dynamic mix of decentralized platforms with certain centralized model. For that, we can see the early version of the Metaverse from the following chapters on DeFi, NFT, and blockchain gaming.

Visa and Mastercard: Tokenomics Going Mainstream

Metaverses are virtual worlds where internet users can socialize, shop, work, and experience various activities. There are many ideas about how the Metaverse will evolve, but some form of cryptocurrencies will likely be the payment of choice. In addition, it is possible to use blockchain to ensure the automation of payment that can fully manage the micropayments required by the Metaverse. For example, we are already seeing people using crypto to buy land and goods in various virtual universes (more examples in the following chapters). (In **Chapter 9**, a three-way competition among public cryptocurrencies, government CBDCs, and Big Tech tokens will be discussed.)

The Metaverse must have a global payment rail – global real-time instantaneous settlement and clearing for payment enabled by cryptocurrency. In Web1.0, we had text images that ran on a desktop computer. Purchasing online was rare and feared. Web2.0 added video and mobile. Buying became one click. Web3.0 will include 3D worlds, virtual, augmented and mixed reality (XR), artificial intelligence and decentralized commerce with a faster adoption rate than previous generations of the internet. Buying will be as instant as a thought.

The reality is that we are still in the early days of tokenomics. Blockchain and crypto transactions are still plagued by serious technical shortcomings, notably scalability and performance. Blockchains cannot, at present, process the huge number of transactions that centralized mobile payment apps can; and for technical reasons, the amount of energy it takes to secure a transaction on the blockchain (POW) increases over time.

But a blockchain-based digital financial future is the vision. New blockchain companies (Ripple, BlockFi, Coinbase, etc.) are democratizing access to traditional financial products (lending, investing, cross-border payments), and giving underbanked populations (especially in the emerging markets) the access to financial tools for the first time. Unsurprisingly, tech-savvy millennials and Gen Z are leading the crypto revolution; according to a CNBC 2021 survey, cryptocurrencies are the only type of investment with disproportionate youth participation: 15 percent of 18- to 34-year-olds own cryptocurrencies, compared to 11 percent of 35- to 64-year-olds, and only 4 percent of those 65 and older.

In 2021, crypto made itself impossible to ignore, with a market cap at times surpassing $3 trillion, partly because cryptocurrencies and unique digital assets have become the outlet of choice for a generation that has been locked out of the traditional financial system. In addition to crypto communities' use of token, what truly brings crypto assets to the mainstream is the increasing involvement of established, major financial institutions like Visa.

Visa has established relationships with crypto exchanges like Coinbase and FTX in the past few years. In 2021, Visa disclosed that it had partnered with no less than 50 crypto platforms via card programs that "make it easy to convert and spend digital currency at 70 million merchants worldwide." Visa, in the fight to stay relevant and keep customers engaged, is achieving user loyalty by making cryptocurrencies an option on their platforms. Working with Anchorage, the first federally chartered digital asset bank and an exclusive Visa digital currency settlement partner, Visa has launched a pilot that allows Crypto.com to send USDC (a kind of "stablecoins," detailed discussion in **Chapter 4**) to Visa to settle a portion of its obligations for the Crypto.com Visa card program.

Visa's standard settlement process requires partners to settle in a traditional fiat currency (government currencies), which can add cost and complexity for businesses built with digital currencies. The ability to settle in USDC can ultimately help Crypto.com and other crypto native companies evaluate fundamentally new business models without the need for traditional fiat currency in their treasury and settlement workflows. (By the way, Visa's treasury upgrades and integration with Anchorage also strengthen Visa's ability to directly support new central bank digital currency (CBDC) as they emerge in the future.)

Visa's competitor Mastercard is also sharpening and adopting its fintech strategy to include and make use of cryptocurrencies. According to a 2021 report from Coindesk, Mastercard announced that it is partnering with five startups to solve global challenges in blockchain as part of Mastercard's "Start Path Crypto" program. The startups that Mastercard chose to partner with are smart-contract builder Ava Labs, AI-focused mobile banking app Envel, peer-to-peer savings platform Kash, bitcoin banking app LVL, and crypto rewards platform NiftyKey.

Similar to Visa's adoption of crypto-based payment systems, Mastercard embraces blockchain and cryptocurrencies in an effort to offer more payment options and exchanges, products and services for its increasingly crypto-curious user base. According to a 2021 Mastercard Newsroom report, Mastercard's philosophy regarding giving users crypto-payment options is not to necessarily encourage, but is more about giving users the "choice" to use cryptocurrencies if they so wish: "Mastercard isn't here to recommend you start using cryptocurrencies. But we are here to enable customers, merchants, and businesses to move digital value – traditional or crypto – however they want. It should be your choice, it's your money."

Mastercard also makes a selling point of the fact that it selects certain cryptocurrencies to join the network, not just opening the floodgates to every cryptocurrency out there. Once a cryptocurrency meets Mastercard's safety, reliability, and risk requirements, it will be considered to be added to the Mastercard network. For the crypto-curious user who is doubtful of the risks involved in dabbling in the sphere, perhaps operating through Mastercard can prove to be a safer experimentation option for newcomers.

Like Visa and Mastercard, many consumer-facing and industrial companies were somewhat late to the game because most blockchain applications were geared toward cryptocurrency or financial transactions from the early years of token economics. But their involvement is increasing as more blockchain business applications, driven by the new metaverse thinking, move from the concept stage to reality. Traditional companies and the public are generally becoming more comfortable with cryptocurrency transactions, which could further increase usage rates.

Metaverse Outlook: Crypto Beyond Currencies

Cryptocurrencies and tokens are transforming not just finance and money but also the ways in which creators can form internet-native organizations to create and share value. Web3 has become a proxy for new economic ideas on how the internet should be architected, and how individuals should share in this value creation – for which the crypto assets will pay an important role.

Crypto assets can be classified into two main categories, according to their principal function: native coins and crypto tokens. Native coins, like bitcoin, generally compete with the traditional forms of money, providing both an alternative currency instrument and a payment infrastructure. Different from native coins, utility tokens are coins that embed some intrinsic values somehow linked to the quality of the issuing entity's business model and to the ecosystem it generates.

In other words, utility token is a sub-class of crypto assets, but not all crypto assets are cryptocurrencies. Crypto assets usually have many of the same features of a cryptocurrency in that there will be a token that serves as a store of value, with the ability to transfer that value but there is usually a second layer of functionality added. Because of the spread of blockchain technology, more crypto assets are coming to the markets.

If the growth of crypto as an asset forced everyone to pay attention to the multitrillion market capitalization of the crypto world in 2021, it is the growth of crypto beyond currencies that has the potential to reverberate across industries. Since 2021, the metaverse community has witnessed the emergence of different crypto sectors, each with different value drivers. In this new era, crypto use cases – various "utility tokens" – unrelated to bitcoin's have finally been validated and achieved meaningful adoption.

Utility tokens will play an active and accelerating role in this new system. A consumer who buys a utility token supports the network stability and liquidity. The more purchases and sales of services or goods happen in the network, the more effective the network will be. The use of utility tokens by new users increases the value of the tokens and consequently the investment value of the other users.

More importantly, the more users in a network, the better the security of that network. That means an investor using the utility token is also providing a better network for another user. Therefore, the distinction between stakeholders will fade: a customer will be an investor, and vice versa. A business company based on utility tokens will potentially benefit from positive feedback where the use of tokens will benefit the overall platform.

For certain payment types, distributed ledger technology is enabling more cost-effective, secure, and in most commercial use cases, fully traceable money movement. In the competitive cross-border payments arena, blockchain enables near-instant and transparent payments, eliminating complex and opaque fee structures. One example of changing business models enabled by token economics is "Tip the Farmer projects," where blockchain tokens enable coffee drinker to trace the original farmer and provide economic value to them directly. (See **Box: Tip the Farmer**.)

In another example, in **Chapter 5** we will see nonfungible tokens (NFTs) bringing new life to the art market first, before subsequently expanding beyond the art market to become the basis for a new economy of culture. The NFT tokens provide a financial infrastructure for the future creator economy, where assets accrue value as communities build around them.

Tip the Farmer

Paramount Software Solution is an example of the "Tip the Farmer" blockchain business model. The company created FarmToPlate.io, a grocery-tracking tool that catalogs the journey of a food item (like an apple) from the farmer who originally planted the seeds, to how it has been transported, to what store the food item made its way to before ending up on the plate of the consumer, all via a QR code. If the customer likes the food, they can choose to micro-tip the farmers. This effort has not only helped save millions of pounds of food waste via its transparency mechanism and other technology features, but it can potentially bring more profit to farmers in developing nations who are struggling economically.

Paramount is not alone in the effort to Tip the Farmer via blockchain. Kahawa 1893, a Kenyan coffee brand with a mission to provide Kenyan farmers with sustainable income, also aims to provide end consumers with a clear picture of where their coffee originates. In 2019, Kahawa 1893 infused blockchain technology into its supply chain and enabled consumers to be able to tip their farmers. Similar to that of Paramount, Kahawa 1893's customers can scan the QR code on Kahawa 1893 merchandise and directly send a tip to the coffee farmer's e-wallet.

According to a 2019 *Forbes* article on the company, the tip travels instantly to the farmer's e-wallet, thanks to Mpesa, Kenya's ubiquitous mobile money provider. The Kenya-based startup BitPesa uses distributed ledger settlement, allowing customers to send and receive low-cost, near-instant payments without a bank account or even an enrolled wallet.

In tandem with these changes, the market for blockchain business applications is heating up as more diverse tokens are emerging, and BaaS software development simplifies blockchain implementation. Diversification means more use cases, and with more use cases comes greater adoption. Because of this positive escalation effect, the crypto industry is branching out. More than just a basket of token assets within a portfolio, crypto is beginning to infiltrate everyday life (for example, financing activities, which will be discussed in the following Chapter 4). That's what Web3 meant to be.

CHAPTER

4

DeFi (Decentralized Finance): Bankless Metaverse

- Fintech 2.0: DeFi vs. CeFi
- Governance Tokens and Revenue Sources
- Stablecoins: Bridging DeFi and CeFi
- Layered Protocols and DeFi Security
- Challenges of DeFi Mass Adoption
- Conquering New Territory: DeFi + NFT, Game, and Social Network

Fintech 2.0: DeFi vs. CeFi

Decentralized finance, or DeFi, is the next frontier in finance for the decentralized Metaverse. DeFi uses a combination of existing blockchain-related technologies – such as digital assets, wallets, smart contracts, and auxiliary services including oracles – to create new forms of financial transactions and bypass inefficient and lumbering traditional institutions like commercial banks. To put it simply, DeFi are financial products like collateralized loans and asset derivatives offered via decentralized blockchain technology, instead of the traditional, centralized systems of banks and exchanges (which can be referred to as CeFi).

In 2020, DeFi took the fintech space by storm by demonstrating the power of self-sovereign and peer to peer finance during pandemic lockdown, government monetary easing policy, and fiscal stimulus policies. According to a recent Statista 2022 report, the number of unique DeFi users grew to about 4.5 million in March 2022, from only 189 users in December 2017. Although the number of users is still small compared with a large commercial bank such as Bank of America that has over 60 million customers, the growth rate during the past two years is phenomenal (see **Figure 4.1**).

DeFi has started to reshape global finance and e-commerce, yet it remains mysterious to most people. Where is the DeFi advantage? As illustrated in **Table 4.1**, DeFi seeks to offer a paradigm shift from the way traditional banking is delivered today.

The key feature of DeFi is that even people without banking accounts can access the modern financial system. According to a World Bank survey in 2017, there are 1.7 billion people globally who do not have access to a bank account. Among these 1.7 billion people, approximately 460 million live in Southeast Asia, 350 million in Africa, and 225 million in China. Even in

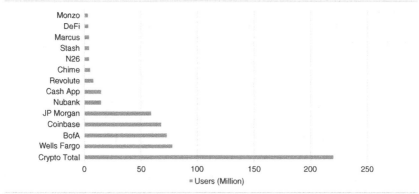

Figure 4.1   DeFi Innovation vs. CeFi Establishment
Source: Adapted from Grayscale Report 2021

Table 4.1   DeFi vs. CeFi

	Centralized Finance (CeFi)	Decentralized Finance (DeFi)
Customers	Restricted to select geographies and privileged customer biases, requiring antidiscrimination laws	Nondiscriminatory equal access for anyone with an internet connection
Structure	Banking offered by traditional companies or legal entities	Banking offered by open-source crypto network software protocols
Participation	Services provided by designated companies and their employees	Services provided peer-to-peer by anyone to anyone else
Ownership	System owned by public or private shareholders or government entities	System owned by public and open to anyone in the user community
Governance	Decisions made by management, industry bodies, and regulators	Decisions made by the protocol, developers, and user community
Asset Custody	Assets held by institutions or custody provider	Assets held directly by users or in noncustodial smart contracts
Unit of Account	Denominated in fiat currency	Denominated in digital asset
Transactions	Executed via intermediaries	Executed via smart contracts
Clearing	Facilitated via clearing house	Facilitated via the protocol
Settlement	3–5 business days, depending on transaction times during Monday to Friday business hours	Seconds to minutes, depending on blockchain with operating times 24 hours per day and 365 days a year
Legal Disputes	Paper legal agreements settled by slow and expensive traditional local court systems	Digital legal agreements settled automatically by software for the cost of a typical transaction fee
Auditability	Authorized third-party audits produced on a quarterly basis	Open-source code and public ledger auditable by anyone on a block-by-block basis
Collateral	Under-to-uncollateralized in many cases with intermediaries; exposing system to risks	Fully-to-overcollateralized in most cases; reducing systemic risks
Risks	Vulnerable to hacks and data breaches	Vulnerable to smart contract hacks and data breaches

Source: Grayscale report 2021

the United States, the most developed capitalist country, there are still about 55 million people who do not have bank accounts.

Now they can become DeFi players easily. Simply with a smartphone and basic internet connectivity, anyone in any corner of the world can access the DeFi protocols and enjoy a wide selection of financial services. For example, decentralized lending services, such as Compound Lending and AAVE, or decentralized asset management services, like MetaMask, imToken, Agent Wallet, Enjin wallet, and more.

Furthermore, DeFi has a few unique advantages based on blockchain technologies:

- **Controllability.** Compared with the CeFi way of handing over assets to financial institutions for custody purposes, DeFi's major advantage is that users always have absolute control over their assets. Additionally, in DeFi, users can play the role of customer or provider in financial transactions at will. They can be either the servicing party or the serviced party, and there is no threshold for entry.
- **Transparency.** The DeFi protocol code is open source and running on a public blockchain, so anyone can verify its security, interaction rules, transaction histories, and actual network usage.
- **Censorship resistance.** Censorship resistance refers to the features of a network that prevent parties from altering or blocking data on it. Once the data is added, it should be virtually impossible to remove or alter it, making it permanent. In DeFi, the blockchain network nodes do not censor and alter individual transactions.
- **Programmability.** DeFi is implemented through smart contracts, and smart contracts can be programmed and adjusted according to functional needs to align with different types of financial logic.
- **Composability.** This is also called *money Lego*, meaning that the base-level DeFi protocols implemented by smart contract can be leveraged by other DeFi projects to compose new DeFi applications. For example, DEX protocol UniSwap and DeFi lending protocol Compound can be leveraged by yield farm protocol to produce optimized yield by lending and swapping assets using these two protocols.

- **Interoperability.** DeFi allows for the tokenization and exchange of the value of digital assets with fiat and physical assets. For example, the Synthetix project is a decentralized synthetic asset issuance protocol built on Ethereum. These synthetic assets are collateralized by the Synthetix Network Token (SNX), which when locked in the contract enables the issuance of synthetic assets (Synths). This pooled collateral model enables users to perform conversions between Synths directly with the smart contract, avoiding the need for counterparties.

Governance Tokens and Revenue Sources

As its name suggests, DeFi is a more decentralized way to conduct transactions. The system is characterized by transparent digital ledgers maintained on multiple computers, so there is no centralized point of failure. Its governance is also decentralized – control rests with the members of a network rather than a central authority. Trust is achieved through public consensus: community members must themselves agree about the validity of transactions, rather than relying on third parties.

As a result, the DeFi transactions can be more secure than traditional finance. For example, in lending and borrowing, DeFi uses overcollaterization and automated liquidation mechanism to mitigate the credit risk of the borrowing party. For derivative trading, smart contracts can be used to get price feeds from various oracle sources and perform real-time margin calls as needed. All these functions are based on DeFi tokens.

DeFi applications are built on top of networks, and each network has its own native tokens. DeFi tokens include mostly two forms of token: governance token and LP (liquidity provider) token. Stablecoins are a special category of tokens and will be discussed separately in the following section.

DeFi Governance Token

DeFi governance tokens allow token holders to govern a blockchain protocol, and in some cases enable them to capture value directly from DeFi application usage. The ideal is like traditional stocks: Individuals can vote on board decisions and profit from dividends when issued. While the traditional finance implementation is standardized, the implementations in DeFi can vary drastically from each other with new models being invented multiple times a year.

The followings are a few samples of governance tokens:

Uniswap (UNI). Uniswap is a trustless and decentralized crypto exchange leveraging smart contracts to provide liquidity and automated market maker (AMM). AMM represents smart contracts that create so-called liquidity pools of tokens, which are automatically traded by an algorithm rather than an order book. AMMs determine token prices based on preset mathematical formulas.

UNI is the governance token for Uniswap protocol. UNI is used for community-led growth, sustainability, and development. To achieve all these, Uniswap enables shared community ownership and a diverse, vibrant, and dedicated governance system. The UNI can be earned by providing liquidity to select pools, which will eventually be used for governance since a larger portion of the supply is issued.

Aave (AAVE). Aave is a decentralized lending system that permits users to lend, borrow, and earn interest on crypto assets without a middleman's services. The Aave protocol runs on the Ethereum blockchain. Aave is a system of smart contracts that allows the assets to be managed by a distributed network of computers running its software.

This simply means that Aave users do not need to trust a person or institution to manage their funds.

They only need to trust that its code will execute as it is written. The AAVE tokens are the native token of Aave. The AAVE token has two key use cases: governance and security. It allows the holders who stake the coin to participate in the operating decisions of the platform.

Maker (MKR): Maker is an Ethereum-based cryptocurrency project responsible for the DAI stablecoin. With the MKR token, users can vote for new changes or make suggestions inside the maker DAO network through the Maker Voting Dashboard.

Compound (COMP): Compound (COMP) token is an ERC 20 token that runs on the Ethereum network. The Compound is a lending protocol. The token holders can vote for suggestions, debates, and implement the network changes through the compound governance dashboard. The token works by allowing the borrowing and lending of a specific set of cryptocurrencies like ETH, DAI, and USDT. Now, any user with those cryptos can lend and borrow crypto instantly without spending the time, effort, and cost of working with a traditional financial intermediary.

Sushi (SUSHI). SUSHI is an ERC-20 token that is issued on the SushiSwap decentralized exchange to the liquidity providers. SushiSwap is a decentralized exchange that allows users to swap cryptocurrency for another similar to UniSwap. The SUSHI token is earned in SushiSwap by providing a liquidity pool, and it can also be staked in exchange for SLP tokens, which are used to govern the protocol. The goal of the SUSHI token is to reward the users of the protocol. To achieve this, the protocol allows users to earn a cut of the SushiSwap fees even when they are no longer providing liquidity to the SushiSwap pool. Users achieve this by staking SUSHI to earn more SUSHI.

DeFi LP (Liquidity Provider) Token

LP tokens represent a crypto liquidity provider's share of a pool, and the crypto liquidity provider remains entirely in control of the token. For example, if you contribute $10 worth of assets to a Balancer pool that has a total worth of $100, you would receive 10 percent of that pool's LP tokens. Holding LP tokens gives liquidity providers complete control over their locked liquidity. Most liquidity pools allow providers to redeem their LP tokens at any time without interference, although many may charge a small penalty if you redeem them too soon.

The relationship between LP tokens and the proportional share of a liquidity pool is used most in two cases: (1) to determine the liquidity provider's share of transaction fees accumulated during the duration of liquidity provision; and (2) to determine how much liquidity is returned to liquidity providers from the liquidity pools when LPs decide to redeem their LP tokens.

Many new use cases for LP tokens are emerging on modern DeFi platforms. These include:

- Staking LP tokens to earn further rewards to incentivize LPs to lock their liquidity into pools. Sometimes, this is called farming.
- Using LP tokens value as a qualifying factor to access initial DEX offering (IDOs) – that is, to participate in certain IDOs, one must hold a certain value of LP tokens.

Following are a few examples of LP tokens used by leading DeFi platforms:

1inch. Crypto liquidity providers using the 1inch DeFi DEX aggregator accrue interest from platform trading fees in the form of the 1INCH token, regardless of the 1inch pool to which they provide liquidity. These 1INCH tokens also serve as the platform's governance token, which means that holding 1INCH tokens comes with proportional voting rights in 1inch's decentralized governance administration.

Uniswap. Uniswap liquidity providers are rewarded with fungible ERC-20 LP tokens, which makes the tokens composable across the broader Ethereum-based DeFi ecosystem. As a result, even though there are generally no direct markets for buying and trading LP tokens themselves, LP tokens like Uniswap's can be used as collateral in lending protocols such as Aave or MakerDAO. It's important to note that Uniswap liquidity provider tokens are not the same as UNI governance tokens, which are used to vote on new proposals and other forms of decentralized decision-making.

SushiSwap. SushiSwap liquidity providers receive ERC-20 SushiSwap Liquidity Provider (SLP) tokens associated with the specific asset they have deposited. For instance, if a user deposits DAI and ETH into a pool, they will receive DAI-ETH SLP tokens. These SLP tokens can then be deposited into a designated DAI-ETH SLP liquidity pool to generate SUSHI, SushiSwap's platform governance token.

Curve. Curve is a decentralized exchange liquidity pool on Ethereum designed for extremely efficient stablecoin trading. Launched in January 2020, Curve allows users to trade between stablecoins with a low-slippage, low-fee algorithm designed specifically for stablecoins and earning fees. Behind the scenes, the tokens held by liquidity pools are also supplied to the Compound protocol or iearn.finance to generate more income for liquidity providers. Liquidity providers stake liquidity to the pool to receive a token-specific LP token rather than an LP token tied to a trading pair.

For instance, if a user lends ETH to the Compound DeFi platform, it is exchanged for a liquidity token called cETH, which automatically accumulates interest for the holder. In addition to allowing Curve's crypto liquidity providers the right to withdraw their ETH plus interest from Compound, Curve users are able to stake their cETH in other liquidity pools to generate passive yields and CRV (Curve's governance token). These LP tokens thereby allow users to achieve an additional layer of utility and potential profits from their initial investment.

Balancer. Balancer is an AMM protocol that enables liquidity pools made up of multiple unevenly weighted assets. Like many of the examples above, Balancer liquidity tokens – called balancer pool tokens (BPT) – are ERC-20 tokens that are composable across the broader Ethereum DeFi ecosystem. However, given Balancer's unique multiasset pool configuration, BPT tokens are underpinned by a basket of crypto assets. Some projects that are built on top of Balancer pools require users to stake BPT tokens to earn rewards.

Kyber Network. Kyber Network aggregates liquidity from a variety of reserves, including token holders, market makers, and DEXs, into a single liquidity pool on its network. Liquidity providers in Kyber's Dynamic Market Maker (DMM) protocol receive DMM LP tokens representing their liquidity pool share. These DMM tokens can then be staked in eligible liquidity mining pools to earn KNC or MATIC (Kyber's and Polygon's respective governance tokens) on top of protocol fees earned through the staking program.

Figure 4.2 shows the total value locked (TVL) in the DeFi ecosystem at the end of March 2022 is about $227 billion. TVL includes all the coins deposited in all of the functions that DeFi protocols offer, including staking, lending, and liquidity pools. The TVL metric is an important gauge of the overall DeFi market assets and liquidity.

In summary, the DeFi protocol and associated DeFi tokens can generate revenue from different sources, such as:

- **Fee dividends.** Protocols may pay fee revenues out to the token holders.

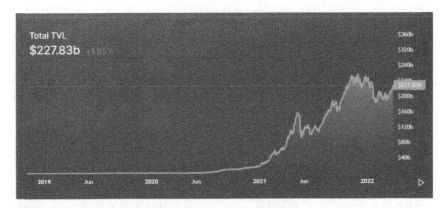

Figure 4.2   Total Value Locked in DeFi System (March 2022)
Source: DeFi Dashboard, https://defillama.com/

- **Token buybacks.** Protocols may use fee revenues to retire (burn) token supply.
- **Token dividends.** Protocols may issue new tokens to groups of holders.
- **Protocol usage.** Protocols may give users fee discounts for holding tokens.
- **Staking rewards.** Examples include ETH2.0 staking or Polkadot's DOT token staking rewards.
- **Governance voting.** Protocols may require tokens for governance voting.

Stablecoins: Bridging DeFi and CeFi

Stablecoins is crucial innovation serving as a catalyst to DeFi adoption. A stablecoin is a class of crypto tokens that attempt to offer price stability and may be backed by a reserve asset. Historically, cryptocurrencies were considered too volatile to facilitate financial transactions other than for speculative trading. The answer to this problem came in the form of "stablecoins," as they attempt to offer the best of both worlds – the real-time clearing and settlement of payments of cryptocurrencies, and the volatility-free stable valuations of fiat currencies.

Stablecoins act as the stable unit of account that allows for more complex financial transactions and derivatives to occur. It points the way toward integrating traditional financial markets with the quickly evolving DeFi industry. As a force for market stability, stablecoins present a primary vehicle for cryptocurrency adoption in loan and credit markets, while inheriting much of the utility previously reserved for only fiat currency.

There are four kinds of stablecoins based on four underlying collateral structures: fiat-backed, crypto-backed, commodity-backed, or commodity-backed. While underlying collateral structures can vary, stablecoins always aim for the same goal: stability.

Fiat Collateral (Off-Chain)

The most popular stablecoins are backed 1:1 by fiat currency. Because the underlying collateral isn't another cryptocurrency, this type of stablecoin is considered an off-chain asset. Fiat collateral remains in reserve with a central issuer or financial institution, and must remain proportionate to the number of stablecoin tokens in circulation.

Some of the biggest stablecoins in this category by market value include Tether (USDT), the Gemini Dollar (GUSD), True USD (TUSD), and Paxos Standard (PAX).

Crypto Collateral (On-Chain)

As the name implies, crypto-collateralized stablecoins are backed by other types of cryptocurrencies as collateral (usually ethereum, bitcoin, and other top cryptocurrencies). This process occurs on-chain and employs smart contracts instead of relying on a central issuer.

When purchasing this kind of stablecoin, you lock your cryptocurrency into a smart contract to obtain tokens of equal representative value. You can then put your stablecoin back into the same smart contract to withdraw your original collateral amount. DAI is the most prominent stablecoin in this category. This is realized by utilizing a collateralized debt position (CDP) via MakerDAO to secure assets as collateral on the blockchain. The term "CDP" is renamed to "Vault" in the recent version of MakerDAO to make it more understandable, because cryptos are often compared to gold.

Crypto-collateralized stablecoins are also overcollateralized to buffer against price fluctuations in the required cryptocurrency collateral asset. For example, if you want to buy $1,000 worth of DAI stablecoins, you may need to deposit $2,000 worth of ETH – this equates to a 200 percent collateralized ratio. If the market price of ETH drops but remains above a set threshold, the excess collateral buffers DAI's price to maintain stability. However, if the ETH price drops below a set threshold, collateral is paid back into the smart contract to liquidate the CDP.

Algorithmic Stablecoins

Algorithmic stablecoins do not use fiat or cryptocurrency as collateral. Instead, their price stability results from the use of specialized algorithms and smart contracts that manage the supply of tokens in circulation. An algorithmic stablecoin system will reduce the number of tokens in circulation when the market price falls below the price of the fiat currency it tracks. Alternatively, if the price of the token exceeds the price of the fiat currency it tracks, new tokens enter into circulation to adjust the stablecoin value downward. Example of such algorithmic stablecoin include Ampleforth or $AMPL token, $FEI, and $FRAX.

Since algorithmic stablecoin does not have collateral, the actual value of such coin will depend on the following factors:

Governance. Many algorithm-based stablecoin protocols supposedly feature a DAO structure. However, only a selected assortment of protocols features an active community responsible for consistently approving improvement proposals. A functional governance smart contract might seem like the ideal choice for ensuring governance in noncollateralized stablecoins. However, it is also important to ensure fair token distribution alongside offering adequate governance privileges to all the stakeholders. Various types of algorithmic stablecoin protocols follow a de facto centralized governance approach.

Incentives. You can find that certain algorithm stablecoin protocols choose the rebase mechanism (rebase is designed in a way that the circulating token supply adjusts automatically according to a token's price fluctuations) in cases where they have to ensure active modification of a number of tokens in a user's wallet. On the other hand, some protocols are aimed at offering returns on alternative investment vehicles such as coupons for removing or adding supply and matching the demand.

Therefore, incentives stand as one of the toughest aspects in determining the efficiency of the best algorithmic stablecoins. Why? You have the instability of the crypto market alongside the fluctuating elements in human psychology and economics. As of now, the only visible incentive with stablecoins is stability.

Token adoption. The factor of token adoption is also an important aspect in defining the ideal choices in an algorithmic stablecoins list. Most protocols are adopted only by a trivial number or a specific limit of DeFi projects. Automated Market Makers, which do not need approval from the partner protocol, are excluded from the adoption. As a result, the usefulness of the algorithm-based tokens takes a dip alongside restricted exposure to new users. The limited levels of token adoption could present limitations for stability on the ground of slower liquidity growth.

Accuracy. Algorithmic stablecoins struggle considerably in maintaining the peg for different complicated reasons. Certain protocols could go out of control to an extent where they are likely to fall in a *death loop*. Therefore, a major protocol needs the mechanism to help the protocol move out of the "death loop."

Commodity-Backed Stablecoins

Commodity-backed stablecoins are collateralized using physical assets like precious metals, oil, and real estate. The most popular commodity to be collateralized is gold; Tether Gold (XAUT) and Paxos Gold (PAXG) are two of the most liquid gold-backed stablecoins. However, it is important to remember that these commodities can, and are more likely to, fluctuate in price and therefore have the potential to lose value.

Commodity-backed stablecoins facilitate investments in assets that may otherwise be out of reach locally. For instance, in many regions, obtaining a gold bar and finding a secure storage location is complex and expensive. As a result, holding physical commodities like gold and silver is not always a realistic proposition.

However, commodity-backed stablecoins also afford utility to those that want to exchange tokens for cash or take possession of the underlying tokenized asset. Holders of Paxos Gold (PAXG) stablecoins can sell them for cash or take possession of the underlying gold. However, because London Good Delivery gold bars range from 370-to-430 per ounce, and each token represents 1 ounce, users must hold a minimum of 430 PAXG to execute token redemption. Once redeemed, token holders can take possession of their gold at vaults throughout the United Kingdom.

Similarly, holders of Tether Gold can redeem XAUT tokens in exchange for physical gold if they complete the TG Commodities Limited verification process and hold a minimum of 430 XAUT. This minimum reflects the standard 430 oz. London Bullion Market Association (LBMA) gold bar. Once XAUT is redeemed, holders can take possession of their gold at a location of their choosing within Switzerland.

Although the ability to redeem gold-backed stablecoins for physical gold is universal across active platforms, other commodity-backed stablecoins lack the same utility. For example, Venezuela's exploratory Petro stablecoin is not redeemable for a barrel of oil. While stablecoins backed by other commodities like real estate have made headlines in recent years, a lack of active projects makes it difficult to draw further comparison.

Layered Protocols and DeFi Security

DeFi's Money Lego Layers

From the technological and functional perspectives, DeFi projects can be divided into different levels of "money Legos" (smart contracts). For example, public chains and protocols belong to the underlying technology, and DEX and lending belong to different levels of applications. Meanwhile, different levels can be combined to build new DeFi applications:

1. **The underlying technologies.** This includes public chains like Ethereum, Solona, Polkadot, and Cosmos, and different wallet applications such as Metamask, Enjin, Agent, imToken, etc.
2. **Stablecoin.** It is a crucial money lego to enable price data and exchange assets. Examples of stablecoin include Tether USDT, USDC, DAI and some versions of algorithm-based stable coins (for example: Basis Cash, Amperforth, Empty Set Dollar, and Frax).
3. **DEX (decentralized exchange)**, such as Uniswap, 0X, and Kyber.
4. **Borrowing and lending protocols**. At present, the most popular application categories on DeFi, MarkerDAO, Compound, AAVE etc. belong to this category.
5. **ETF, synthetic assets, ABS (asset-backed security tokens) and other financial derivatives.** Not only are the derivatives able to be expressed in the form of digital assets, but also physical assets can be expressed in the form of digital assets.
6. **DeFi index funds, insurance products, and supply chain finance.** At present, DeFi is not mature enough to support sophisticated application in these areas, but innovation is rapidly emerging, especially in the insurance market.

It should be noted that from the second to the sixth level, the assistance of blockchain oracles (oracles connect block-chains to external systems) is required to provide authentic data. When different levels are combined, pay close attention to associated security issues. Following is a list of most widely exploited security vulnerabilities in the DeFi systems.

Incorrect Liquidity Pool Calculations

When the value of tokens within the DeFi pool is priced, one of the most exploited forms of attack in DeFi smart contracts is the manipulation of price by hackers.

DeFi participants gain a stake of tokens when they invest in a pool, allowing them to extract value in the future. Rather than using an external oracle, some poorly designed DeFi protocol frequently computes the value of the tokens they hold depend-ing on the present composition of the pool. Flash loan assaults take advantage of this by dramatically unbalancing a pool for the period of a transaction. This uneven pool causes the token's value to be calculated incorrectly, allowing the attacker to drain value from the pool. (Belt Finance, Rari Capital, and BurgerSwap are examples of protocols that have been exploited in this fashion.)

Stolen and Leaked Private Keys

To manage the access to and control over blockchain accounts, blockchain technologies employ public key cryptography. The address of a blockchain account is derived from a public key that is connected to a private key. Any transaction conducted on that account's behalf must be digitally signed with the cor-rect private key. As a result, private keys are the focus of numer-ous blockchain attacks. Keys can be stolen or compromised in a variety of ways, including:

- **Compromised cryptocurrency wallets (like MetaMask).** MetaMask is commonly used to interact with and per-form transactions on the Ethereum blockchain. Several DeFi users and projects – including the CEO of Nexus

Mutual and the EasyFi project – have lost cryptocurrency when they used malicious versions of MetaMask installed on their machines.

- **Leaked/stolen mnemonic phrase.** Mnemonic phrases are a common way to make private keys easier to remember or enter when recovering or setting up a new wallet. Some DeFi-related hacks have involved the theft or accidental exposure of these keys.
- **Poor key generation.** Private keys should be generated using a secure random number generator. If these keys are generated improperly with a poor source of randomness, then an attacker may be able to guess them and gain control over a blockchain account.
- **Phishing attacks on the private key.** Hackers usually target developers or top executives from prominent DeFi protocols to phish for private key. For example, in November 2021, the private key of BZX protocol's developer was stolen and $55 million lost. This attack granted the hacker access to the content of the bZx developer's wallet and the private keys to the BSC and Polygon deployment of bZx Protocol. For this kind of attack, it can be easily avoided by separating the development duties from operation duties. The operation team should hold multiple signature keys and deploy protocol contracts using a safe workstation.

Poor Access Control of Privileged Functionalities

Privileged functionalities are common in every DeFi smart contract. These functions are only to be called by the contract's owner, and access control must be in place to ensure that. Access is usually controlled by specifying that calls to a function must be made by one or more addresses from a list of addresses. These access controls are sometimes omitted or built in a way that allows an attacker to circumvent them. If this happens, the attacker gains privileged access to the contract, allowing them to extract value from it.

The Poly Network and Punk Protocol hacks are two recent examples of such a vulnerability, in which the attacker claimed ownership over the projects' contracts and utilized that access to drain value from them.

Frontrunning Attacks

Blockchains do not immediately add transactions to the distributed ledger. Transactions are broadcasted to the blockchain network as soon as they are created, but they are stored in mempools on each blockchain node until they are added to the ledger as part of a block.

The gap between the creation of a transaction and its inclusion in the ledger creates the opportunity for frontrunning attacks. The attackers, commonly an autonomous program (bot), will look for transactions that they can exploit. If they see one, they create their own version of the transaction with a higher transaction fee and transmit it to the network. Since blockchain miners commonly order transactions in blocks based on their transaction fee, the attacker's transaction comes before the original one, netting them a profit.

Frontrunning impacts DeFi security in different ways. Many bots will use frontrunning to make a profit based on foreknowledge of users' transactions. In some cases, this is malicious, while in others (such as the DODO DEX and Punk Protocol hacks), a bot front-runs an attempted exploit and then returns the stolen tokens to the exploited protocol.

Rug Pulls

Many attacks on DeFi protocols come from external threats, but this is not always true. Rug pulls are attacks committed by the owners and developers of the DeFi protocols themselves. In a rug pull scheme, someone inside the company with privileged access to its contracts uses this access to drain value from the protocol. Typically, the project and the team behind it then disappear, leaving the victims with little recourse to address the issue.

One of the Chainalysis blog post noted that rug pulls have emerged as the go-to scam of the DeFi ecosystem, accounting for 37 percent of all cryptocurrency scam revenue in 2021, versus just 1 percent in 2020. The Chainalysis blog post also provides examples of some of the biggest rug pulls of 2021. (See **Box: AnubisDAO and Up1 Rug Pull**.) It is important for

AnubisDAO and Up1 Rug Pull

The AnubisDAO case was selected by a Chainalysis 2021 report as one of the biggest rug pull of this year, with over $58 million worth of cryptocurrency stolen. According to the post, AnubisDAO was launched on October 28, 2021, with claims of offering a decentralized currency backed by *several* assets. However, the project didn't contain a website or white paper, and all the developers went by pseudonyms. Miraculously, AnubisDAO still managed to raise nearly $60 million overnight. Yet 20 hours later, all of those funds disappeared from AnubisDAO's liquidity pool.

While AnubisDAO demonstrates a large-scale DeFi rug pull, new cases are occurring *almost* daily. An early Ethereum and DeFi investor who wishes to remain anonymous told Cointelegraph that they fell victim to a rug pull on December 19, 2021. The anonymous source shared that the project is called "up1.network," noting that many early Ethereum investors were discussing Up1 in a Discord chat group. The anonymous source shared that the project is called "up1.network," noting that many early Ethereum investors were discussing Up1 in a Discord chat group. According to the victims:

> People I trusted were mentioning the project, so I checked it out. I thought it was strange to see Up1 giving away airdrops, but thought it could have been affiliated with a DeFi token I had. I then connected my MetaMask wallet and clicked on 'get airdrop' but kept getting an error message. I did this three times, which gave the project access to my account.

Unfortunately, once Up1 gained access to their account, three DeFi tokens worth $50,000 were instantly taken. "I revoked access after the fact on Etherum so they couldn't steal any more tokens," they mentioned. The Ethereum investor then checked the DeFi platform Zerion where they saw the notifications that the DeFi tokens had left their wallet.

And Zerion also provided them with a wallet address to where the funds went, along with a message: "0xc28a580acc42294787f44cffbaa788eaa4958056; *You gave a web3 site / smart contract unlimited access to your funds (check who you gave access to and revoke here).*"

DeFi user to click on approve or other messages prompted by a dialog from a web3 app using their own wallets such as Metamask, Enjin, or Agent wallet as the hacker's contract may be able to gain access to all your tokens and then withdraw them from the affected wallet address.

DeFi Cross-Chain Bridge Attack

In August 2021, Poly Network, an important cross-chain protocol for swapping tokens across multiple blockchains, suffered an attack. As a result, more than $600 million was stolen, which was the largest DeFi hack. The day after the attack, the hacker posted a series of Q&As via memos in Ethereum transactions. He said the attack was "for fun :)," and because "cross-chain hacking is hot." This incident truly highlighted the importance of securing cross-chain protocols.

In a way, the high frequency and large monetary impact of these cross-chain attacks is a consequence of the growing enthusiasm for cross-chain protocols. The emergence of cross-chain DEXs has made it simple and fast for DeFi users to trade assets across different chains. In a traditional centralized exchange (CEX), cross-chain transactions take tens of minutes or even several days to process and can involve large fees. Thus, a cross-chain DEX (which only takes two to three steps and a few seconds to complete a cross-chain transaction) can offer improved user experiences. However, the recent and frequent security incidents surrounding cross-chain protocols indicates that, without strong security guarantees, better speed, efficacy, and rates might not be enough.

Following is a brief overview of the potential risks of cross-chain DeFi transactions, with specific examples of past attacks.

1. Fake Deposit – THORChain

This exploit can happen when the protocol has some vulnerability in the contracts that enables hackers to trick the network into thinking they have deposited funds into the contract, when they have not. This enables the attacker to ask for a refund of a deposit that was never there in the first place.

THORChain suffered this type of attack. The function for verifying the token address checked to see if the ticker of the received token was ETH. The attacker forged an ERC-20 token with the ticker ETH to deceive the cross-chain bridge and then received real ETH tokens as a "refund" to themselves.

2. Multi-Signature and Quota Vulnerabilities – ChainSwap

ChainSwap funds were stolen for a second time because the quotas for whitelist addresses of cross-chain bridges were increased by a node. When the whitelist was checked, the multi-signature originally required became a single signature due to a configuration error. The attacker only needed one of the signatures to transfer the assets by calling the Receive function on the other chain.

3. Redemption Risk – Anyswap

To explain this, we take the BNB of OKExChain (OEC) as an example. Anyswap issued BNB on OEC without official authorization from Binance. The BNB on OEC is not equivalent to the BNB (the cryptocurrency issued by the Binance exchange) on Binance Smart Chain (BSC). It is a BNB bond issued by Anyswap and circulated on OEC.

4. Private Key Leakage – Anyswap and Axie Infinity

The main reason that Anyswap was attacked recently was that the signature used a repeated R-value. If two transactions signed by the same account have the same R-value of the RSV signature, the hacker can reversely derive the private key of the account. Since the account could be used on BSC, ETH, and FTM, the assets on multiple chains of the account were stolen. The excessive authority of a single account is also a risk factor.

Another example is a nearly $650 million hack from the Axie Infinity NFT game in March 2022. The hacker was able to steal five validators' private keys to approve and validate cross-chain transactions from Ronin network to Ethereum network. (Ronin blockchain network is Axie Infinity's private blockchain, which has nine validators, but only need five validators to approve cross transactions.)

5. Flash Loan Arbitrage – bEarn Fi
The attacker lent BUSD through flash loans, generated ibBUSD through Alpaca Finance's lending, and then exploited bEarn Fi's contract strategy vulnerabilities to carry out arbitrage attacks between ibBUSD and BUSD.

6. Access Control of Contract – Poly Network
By disguising a transaction to replace the Keeper (relayer) as a normal cross-chain transaction, the attacker replaced the address of the Keeper in the Poly Network with his own address. The verification function for relayer was not secure enough, and it only checked 4 bytes. The attacker found 4 bytes that met the requirements and successfully converted his own address into the Keeper address and signed it. Then, the hacker called the LockProxy contract and looted all the cross-chain assets of Poly Network.

7: Signature Verification Bypass Due to Supply Chain Code Risk – Wormhole
Wormhole – a web-based blockchain "bridge" that enables users to convert cryptocurrencies was attacked and over $320 million was stolen in February 2022. The attack happened because Wormhole Protocol code uses a third-party library code to verify the depositor's signature. The hacker was able to use a malicious transaction to bypass the deprecated signature verification algorithm and stole 120,000 Ethereum coins.

Bumpy Road to Mass Adoption

While many believe that DeFi will be the foundation of future financial systems, the widespread adoption by retail consumers has not taken place. Most of the current DeFi transactions are from "digital asset native" institutions and traders. Even within the crypto community, only a small percentage of cryptocurrency players are active in the DeFi field. This seems to suggest that there is a significant amount of growth left to come in DeFi. (See, again, **Figure 4.1.**)

But there are also many challenges to widespread DeFi adoption. For most people, DeFi is too technical, too volatile, and too "geeky" to understand and use. On one hand, there are many nontech hurdles relating to the market conditions, such as:

- **User education needed.** Because of the complex and composable logic of DeFi protocol, a huge amount of technical background is required to get comfortable transacting with DeFi protocols.
- **Regulatory uncertainty.** The US regulators, especially the SEC, are developing DeFi-related regulations. For example, should certain DeFi tokens be deemed securities? (See detailed discussion in Chapter 9.)
- **Lack of industry standards.** "10 Fundamental Rights for Crypto Users" proposed by Binance, the world biggest crypto exchange, is a good start. (See **Box: Crypto User's Bill of Rights.**)
- **Lack of global reach.** It's not easy for DeFi builders to reach global markets and build a global community.

On the other hand, there are several blockchain and DeFi-specific challenges, as crypto networks and DeFi technologies are still nascent. Several aspects, such as underlying network scalability, KYC/AML, and UI/UX design, still need to improve before DeFi can service a more sizable global financial market.

Crypto User's Bill of Rights

Binance, the world's biggest crypto exchange, has proposed crypto users' "Bill of Rights." The bill of rights can be applied to the DeFi as well. Binance's bill of rights includes calls for privacy and the right to use DeFi tools. The manifesto-like document calls for universal access to financial tools, strict protections for personal data, and other measures.

Touting the document as a global regulatory framework for crypto markets, Binance also created provisions directed at exchanges – "10 Fundamental Rights for Crypto Users." Those provisions covered the obligations of exchange to protect users from bad actors and ensure enough liquidity for frictionless trading. Unless noted, most of the following articles apply to DeFi. (The notion of "financial tools" in the original text certainly covers "DeFi tools." For the provisions that only apply to centralized exchanges (articles 3 and 5) and do not apply to DeFi, we have provided comments.)

10 Fundamental Rights for Crypto Users

I. Every human being should have access to financial tools, like crypto, that allow for greater economic independence.

II. Industry participants have a responsibility to work with regulators and policymakers to shape new standards for crypto assets. Smart regulation encourages innovation and helps keep users safe.

III. **(This article does** not **apply to DeFi tool; it only applies to centralized exchange since DeFi tools cannot provide KYC.)**

Responsible crypto platforms have an obligation to protect users from bad actors and implement Know Your Customer (KYC) processes to prevent financial crimes.

IV. Privacy is a human right, and personally identifiable information (PII) data should be subject to strict levels of protection.

V. **(This article does not apply to DeFi since the DeFi users keep their own funds.)**

Crypto users have the right to access exchanges that keep their funds secure, in safe custody with comprehensive deposit insurance.

VI. Healthy markets should maintain a robust level of liquidity to ensure a stable and frictionless trading environment.

VII. Regulation and innovation are not mutually exclusive. Crypto users deserve safe access to emerging technologies and practices, including NFTs, stablecoins, staking, yield-farming, and more.

VIII. Closing the knowledge gap is essential when it comes to crypto. Users have the right to accurate information on crypto assets, without fear of falling victim to unfair or deceptive advertising.

IX. Marketplaces that offer derivative instruments should be subject to the appropriate regulations. This ensures that all users meet eligibility requirements and that their transactions are fairly settled.

X. Crypto regulation is inevitable. Users have the right to share their voice on how the industry should evolve with their blockchain platform of choice.

Scalable Base Blockchain

One of the main obstacles to mainstream adoption of DeFi is the scalability of the base blockchain. The cost to engage in DeFi transactions in Ethereum is very expensive, which could be multiple hundreds of dollars for lending your token or borrowing tokens, and the transaction per second is slow at an average 15 per second. (There are many scalability solutions, as discussed in Chapter 2.)

The most promising solution is the layer 2 protocol such as roll-up protocol on Ethereum or some other layer 1 approach such as Binance Smart Chain, Solana, or Terra. The cross-chain solutions such as Polkadot and Cosmos are also attractive for DeFi protocol developers to bridge different DeFi protocols to form new DeFi systems. It is important to note the "blockchain trilemma," which states that it is impossible to achieve secure, scalable, and decentralization at the same time. One can only meet two of the three requirements at the same time.

For example, you must sacrifice the decentralization if you need to meet security and scalability requirements. Or you have to compromise with scalability if you demand security and decentralization. The Ethereum base chain has strong security and decentralization, but that comes at the expense of scalability. Meanwhile, all layer 2 solutions or other layer 1 alternatives to Ethereum have been able to achieve good scalability – at the expense of decentralization.

KYC/AML

Because of the peer-to-peer nature of DeFi businesses, DeFi transactions involving natively digital assets may be difficult to regulate through traditional AML/KYC (anti-money laundering/know your customers) controls, since users are pseudonymous by default, transactions are resistant to blockage, assets are resistant to seizure, and many transactions involve noncustodial wallets not directly tied to individuals.

For example, if an individual provides liquidity to a DeFi exchange pair pool and then cashes out by withdrawing their liquidity, how will it be possible for a DeFi protocol developer to know who exactly the liquidity providers are? If a user leverages DeFi to lend or borrow crypto currency using a self-sovereign wallet via smart contract, how can the DeFi lending protocol know the user's actual identity? If DeFi platforms do not know the user's actual identity, how can they provide accurate reporting? If a user stakes a DeFi token via a smart contract and gets the rewards for staking and protecting the security of the DeFi platform or the base blockchain, how does the anonymous nature of staking meet the reporting requirements?

One promising solution is to leverage Decentralized Identity (DID) standard and Verifiable Claim standards currently under heavy development at W3C and Decentralized Identity foundation. As the DID standard matures, KYC/AML can be applied to DeFi protocols such that the DeFi protocols are integrated with "verifiable credential" providers to implement the required AML / KYC process for some transaction while simultaneously protecting users' privacy using zero knowledge proof (more about this in the Chapter 7).

Another possible approach is something already implemented by China's sovereign digital currency (e-CNY; see detailed discussion in Chapter 9). The e-CNY allows smart amounts to be transferred between individuals without KYC/AML. Furthermore, Big Data and AI can be used to track terrorist financing and money-laundering activities when there is no KYC/AML.

UI/UX Design

Most DeFi applications developed so far have very poor UI/UX design. UI (user interface) and UX (user experience) mean very different things, but only together can they help create a user-friendly DeFi product for mainstream adoption. UI is directly related to how to present your product, whereas UX to the overall interaction between the app and the user. If you want your DeFi products to stand out from the crowded market, you need an innovative UI design, but UI is only a small fraction of the ecosystem that generates and influences the UX of a product.

UI Design involves managing and manipulating various forms of content (text, images, graphics, video), fields, functions (buttons, labels, boxes, commands, drop-down lists), and action items (what happens when you click on certain links). UI requires a touch of art and emotion, and the goal is to create an engaging interface to make the DeFi products attractive to users. UX Design has a much wider spectrum and includes several more elements (architecture, interaction, content, user research), which are integrated together to meet the needs of DeFi users. Currently, the UX side has a lot to improve to be user-friendly to the broad population.

The most important catalyst for mass adoption, interestingly, may come from CeFi, the apparent nemesis of DeFi. Because of their much broader customer reach, CeFi's participation will be very important for DeFi to be adopted by mainstream investors. Indeed, the blue-chip financial institutions such as Fidelity, JPMorgan Chase, and Goldman Sachs have all been trying to offer some degree of DeFi services to their customers (see **Figure 4.3**).

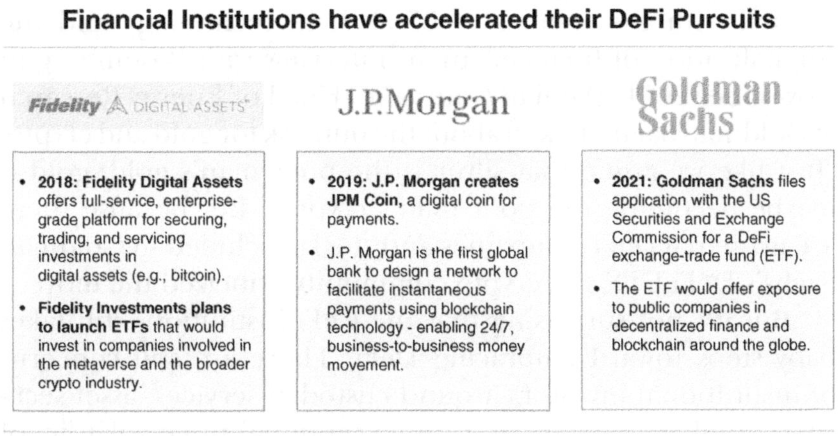

Financial Institutions have accelerated their DeFi Pursuits

Fidelity DIGITAL ASSETS	J.P.Morgan	Goldman Sachs
• **2018: Fidelity Digital Assets** offers full-service, enterprise-grade platform for securing, trading, and servicing investments in digital assets (e.g., bitcoin). • **Fidelity Investments plans to launch ETFs** that would invest in companies involved in the metaverse and the broader crypto industry.	• **2019: J.P. Morgan creates JPM Coin,** a digital coin for payments. • J.P. Morgan is the first global bank to design a network to facilitate instantaneous payments using blockchain technology - enabling 24/7, business-to-business money movement.	• **2021: Goldman Sachs** files application with the US Securities and Exchange Commission for a DeFi exchange-trade fund (ETF). • The ETF would offer exposure to public companies in decentralized finance and blockchain around the globe.

Figure 4.3   Wall Street Embraces DeFi
Source: JPMorgan, Reuters, Blockworks, CoinDesk

Since October 2018, asset manager Fidelity's Digital Asset Services has started to provide custody for cryptocurrencies such as bitcoin and to execute trades on multiple exchanges for investors such as hedge funds and family offices. When reiterating Fidelity's focus on connecting the crypto assets industry with traditional financial sectors recently, Fidelity CEO Abby Johnson called the company's crypto custody business "incredibly successful."

In July 2021, investment bank JPMorgan Chase has reportedly green-lighted its advisors to provide clients with access to five cryptocurrency funds. Four of them are from Grayscale Investments – The Bitcoin Trust, Bitcoin Cash Trust, Ethereum Trust, and Ethereum Classic Trust, and the fifth is Osprey Funds' Bitcoin Trust. The five funds are approved for all JPMorgan's wealth management clients seeking investment advice, and JPMorgan advisors can take orders to buy and sell five cryptocurrency products. These five funds can be viewed as a bridge to DeFi funds, although they are not DeFi funds.

For Goldman Sachs, its DeFi ETF funds filed with SEC in July 2021 would provide investors with exposure to companies aligned with the themes of blockchain technology and the "digitalization of finance." In an interview with Bloomberg in November 2021, Damian Courvalin, Head of Energy Research at Goldman Sachs, talked about the outlook for gold and crypto: "Just like we argue that silver is the poor man's gold, gold is maybe becoming the poor man's crypto." But because there are no actual DeFi companies and stocks included in Goldman Sachs's DeFi ETF, the crypto community criticized the move.

But it's not surprising that the CeFi institutions still takes baby steps toward embracing DeFi. There are still concerns of institutional investors around custodial services, asset security, compliance issues, and the exceptional price volatility of crypto assets. Furthermore, many DeFi projects use governance tokens affiliated with the protocol, but it remains to be seen whether they will accrue long-run sustainable value tied to the fundamental growth of such DApp projects. Still the big trend is clear: How CeFi institutions respond to DeFi will impact their role in the coming evolution of the global financial system, and their inevitable embracing of DeFi will accelerate the spread DeFi significantly.

Conquering New Territory: DeFi + NFT, Game, and Social Network

In addition to competing and collaborating with CeFi in traditional finance, DeFi is expanding into completely new territories of decentralized finance, as illustrated below by its integration with NFTs, games, and social networks.

DeFi and NFT

DeFi and NFT can be leveraged together in a blockchain project to provide the following values:

> **Provide financing and liquidity for NFT project.** Suppose a painting is worth $1 million, yet it does not have any value unless somebody is willing to pay for it. In this situation, DeFi and NFTs do combined work to solve this issue. One

approach is to use NFT collectibles and art as collateral against DeFi loans to provide liquidity for the project. As traditional arts have been used as collateral in the real world since the beginning, moving it to the crypto world and NFT art appears to be a logical step forward. Another approach is to use NFTs to tokenize the artwork and then fractionalize the NFT token into smaller portions using DeFi smart contract, so that more people can join to provide the liquidity to purchase the artwork.

NFT Ownership and DeFi. With NFT ownership mechanism, such as using NFT to represent the digital rights of a song, the NFT owner (the musician) can use that NFT as collateral in DeFi to get loans so the musician can start his own recording business. Or he can stake it in a DeFi platform to gain staking incomes from the DeFi ecosystem in addition to the royalty from NFT ownership. (See in-depth NFT discussion in Chapter 5.)

Solving Bonding Curve Issue

Some DeFi protocols introduced a "bonding curve" to distribute liquidity across the entire curve to encourage early participation into the DeFi ecosystem. (A bonding curve is a mathematical formula used to set a relationship between a token's price and its supply.) The problem with the bonding curve is that when more people are taking part in the system, the rewards get diminished. With the assistance of NFT, one can introduce a mechanism to select desired custom price sizes for liquidity providers, which is independent of the bonding curve and unique to the NFT holders.

Game + DeFi = GameFi

GameFi is the combination of video games (gaming) and decentralized finance (DeFi). The technology used for this type of video game is blockchain technology, which allows players to be the sole and verified owners of the virtual elements of the

Pay-to-win vs. Play-to-earn

Traditionally, gamers must pay to gain an advantage, such as upgrading, reducing waiting time, or buying a virtual object. So, the games are costly to the players (pay-to-win). In contrast, GameFi may generate income for gamers when they play their favorite games (play-to-earn, or P2E).

But first, the gamers need to have blockchain equipment for P2E. Most games require the following steps to start playing:

1. **Create a crypto wallet.** Unlike traditional games, in which you need a username and password, games that use blockchain technology require you to identify yourself using a wallet (like MetaMask). Depending on the game you will need one type of wallet or another. For example, the famous game Axie Infinity requires a wallet compatible with Ethereum.
2. **Add funds to your wallet.** In order to complete step 3, you will need to add funds to your wallet. Depending on the game, you will need to use one cryptocurrency or a few. The Cryptoblades game, for example, requires users to download MetaMask, purchase Binance (BNB) currency, and exchange it for the game's native cryptocurrency, SKILL.
3. **Buy the basic digital assets to start.** To generate profits in most GameFi games, you need to do so through your avatar or similar digital assets. This means that before playing, you will need to buy them. For example, Axie Infinity requires its players to have three Axies in their wallet to start playing. What's truly interesting is that decentralized autonomous organizations (DAOs) have emerged to help blockchain gamers to finance the initial purchase of digital assets. (See case studies in Chapter 10).

game. In traditional video games, the predominant model is "pay-to-win," in which players must pay to gain an advantage, such as upgrading, reducing waiting time, or buying a virtual object. GameFi, on the other hand, introduces the play-to-earn (P2E) model (see **Box: Pay-to-win v. Play-to-earn**). (See in-depth Blockchain Gaming discussion in Chapter 6.)

Emergence of SocialFi

Social finance (SocialFi) is the latest development in the social media marketplace that combines social media platforms, Web 3, NFTs, and DeFi. These new social networks have been created to provide rewards and benefits to users through tokenizing social influence. The idea of SocialFi began back in 2017. However, it was not until late in 2021 that the idea began gaining traction, following users' better understanding of DeFi.

Notably, SocialFi is building a completely decentralized and self-consistent economic system. SocialFi seeks to give control back to their users over how they interact with others on the platforms. In the new concept, users own and manage the platform and creative content. This means that gone are the days when the likes of Facebook, Instagram, and Twitter had total control over social interactions. The buzz around SocialFi has created high expectations among communities calling it the "new outlet."

In the Metaverse, all the above plus upcoming DeFi innovations, can be summarized as "MetaFi" – the financial innovation in the Metaverse that powers the creator economy, which will be discussed at the end of this book. Many DeFi players today are still more or less centralized, and soon the more decentralized DeFi players will play major roles in metaverse applications (see **Figure 4.4**).

To conclude, the current DeFi market is simultaneously both extremely promising and incredibly immature. While it's

Traditional, Crypto, and Decentralized Finance Service Providers

	Decentralized	Centralized (Crypto)	Centralized (Traditional Finance)
Stablecoins	MAKER	tether	
Lending and Borrowing	Compound	BlockFi	ROCKET Mortgage by Quicken Loans
Exchanges	UNISWAP	coinbase	Robinhood
Derivatives	SYNTHETIX	Deribit	CME Group
Data	Chainlink	COINMETRICS	Bloomberg
Asset Management	Yearn.Finance	GRAYSCALE	BlackRock

Figure 4.4   Decentralized DeFi Players Emerging
Source: Grayscale Report, 2021

too early to see the widespread adoption of DeFi and Web3, the latest integration among DeFi, NFT, gaming, and social work indicates that more tech innovation and more user cases are coming, and with more user cases comes greater adoption. More than just a basket of assets within a portfolio, crypto is beginning to infiltrate everyday life through DeFi applications.

Web3 has become a proxy for new economic ideas on how the internet should be architected, and how individuals should share in this value creation. In that context, DeFi applications are not just transforming finance and money, but also the ways in which creators can form digital native organizations to create and share value. In the following chapter, we turn our attention to NFTs and games, which are expanding the blockchain industry out of its financial origin into much broader sectors in the digital economy.

CHAPTER 5

NFTs, Creator Economy, and Open Metaverse

- 2021 – The Year of NFT
- Co-Evolution of Art and Tech
- NFTs and Generative Art
- Creator Economy: Beyond the Bored Apes
- Going Mainstream with Brands and Fashion
- Challenges to the NFTs Metaverse

2021 – The Year of the NFT

Without a doubt, 2021 was the year of the nonfungible token (NFT), unique tokenized representations of digital files that are exchanged on public blockchains. NFT rose from obscurity to front-page news, generating digital assets to represent every possible real-world object, from art and music to tacos and toilet paper. From the advent of novel mechanisms for increasing the liquidity of digital goods and sparking conversations globally, you can barely visit an online space without reading about NFTs.

An NFT is a token on blockchains that contains unique metadata that differentiates it from other tokens. While cryptocurrency is fungible and can be easily transferred, NFTs are not fungible, meaning that each NFT is unique and not

Table 5.1   Differences between Fungible and Non-Fungible Tokens

Criteria	Fungible Tokens (FT)	Nonfungible Tokens (NFT)
Interchangeability	Fungible tokens are interchangeable, but there is no additional value associated with such interchange.	Nonfungible tokens are not interchangeable, as each of them represents unique assets.
Value Transfer	Value transfer depends on the number of tokens in the ownership of a person.	The value of the unique asset represented by NFT is helpful in their value transfer.
Divisibility	Fungible tokens can be divided into smaller parts, and the smaller parts can help in paying off the larger sums.	NFTs are not divisible and have their value as a whole entity.
Token standards	Fungible tokens depend on the ERC-20 standard.	Non-fungible tokens leverage the ERC-721 standard.

interchangeable with another NFT. In other words, while one bitcoin is equivalent to another bitcoin, no two NFTs are the same. (See **Table 5.1** for a comparison of cryptocurrency and NFT.)

As such, NFTs can be used to store much more complex and individual-specific information (e.g., rare collectibles like "digitally signed copy of the first tweet by Elon Musk"), among others. Government documents such as marriage certificates, land registrars, food-grade ratings, and driver's licenses can all be tokenized using NFTs. In retail, consumers can use NFTs to verify the legitimacy of luxury goods. When companies like Taco Bell announce they are launching NFTs, it's clear that something special is happening in the blockchain world.

Interestingly, it's the artist community – arguably the furthest from the blockchain tech world – that started the NFT boom. NFTs represent provable rights for scarce digital art, and they are taking the world by storm. Meanwhile, artists around the world are finding that they can now monetize their creative works at far better rates than in the traditional world through selling NFTs. By December 2021, collections like Cryptopunks

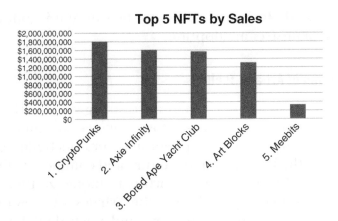

Figure 5.1   Major NFT Projects (by the end of 2021)
Source: Blockchain101.com; nonfungible.com

and Bored Ape Yacht Club (BAYC) had become household names, while celebrity artists like Beeple sold some of their digital creations for more than $69 million, and the global market reached over $40 billion in sales. **(See Figure 5.1: Major NFT Projects.)**

The NFTs were more than just digital photos in jpegs, too; verticals like gaming, music, and sports started to integrate NFTs into their ecosystems, nurturing new projects like Axie Infinity (a "play-to-own" game, detailed discussion in **Chapter 6**). Altogether, 2021 cemented the digital primitive's demand for verifiable ownership, and the NFT market continues to capture value at an incredible pace.

No doubt, NFTs are playing a major role in bringing blockchain to the mainstream. NFTs provide average internet users an easy entry point to the blockchain, even with the high transaction fees (which will be covered later in the chapter). Today, people around the globe have been minting their own NFTs representing art, photography, music, collectibles, in-game items, and almost everything else. In short, the Web3 and Metaverse would not be where it is today without NFT mainstream adoption.

Co-Evolution of Art and Tech

NFTs have led to a digital renaissance taking place in the world of art and content creation – a flourishing ecosystem of artists, musicians, writers, photographers, and curators building communities at the intersection of culture and commerce. NFT Art offers many things – a way for artists to monetize their work, a method for big crypto players (the "whales") to reward new artists like the patrons of classic art, and a new digital medium for creating art itself.

But why now? Let's review the development of art and technology innovation in human history since the Renaissance. See **Table 5.2** for the different stages of art history alongside major technological milestones.

Table 5.2   Co-Evolution of Tech and Art

Four Phases of Art and Tech	Technologies That Influenced Art in Recent Two Centuries	Influences on Art
Manual Craft	- Synthetic pigment created via modern chemical technology - Artificial lighting technology	- Synthetic pigments promote modern art - Modern lighting and cultural relic protection technologies promote the development of museums
Image Printing	- Photography technology - Modern printing tech after photoengraving	- Photography art emerges, promoting modernism - Large-scale, accurate image reproduction and dissemination emerges
Mass Media	- Movie technology - Radio, TV, and video technology	- Emergence of film art - Prosperity of "pop art" promoted by radio and television
Digital Art	- Computer tech - Digital display tech (incl. AR/VR) - Internet/AI/ blockchain	- Brand new medium and methods for art creation - New way of art dissemination

Phase 1: Manual Craft

The first stage could be referred to as the *manual craft stage*. This stage began in prehistoric times and continues to this day. At this stage, the artist uses a manual way to create artworks, which may be a painting, a sculpture, or calligraphy.

At this stage, the artwork created is unique. Appreciation of artworks emphasized on gazing and contemplating in front of the "original objects." Collectors can use buy artworks to acquire the related property rights and be in the presence of artworks day and night. It can be said that the core institutions in today's art ecosystem, such as museums, galleries, art fairs, and auction houses, are all products of this manual craft stage. For this stage, the property right and physical possession of artworks is the key.

Phase 2: Image Printing

Daguerreotype photography was a landmark invention in 1839. Since its invention, photography as a new art category has emerged. The combination of photography and other technologies has spawned many new inventions: for example, the combination of photography and printing gave birth to photo engraving and printing technology. Since then, it has become possible to replicate images accurately for numerous copies. The combination of photography and other technologies has created many new forms of art. Since then, for many customers, one can obtain a reliable image copy at a very low cost. This change has transformed art from being of exclusive use to few people to a resource that the public can obtain, which greatly increases the popularity of art and changes the behavior of art appreciation or art consumption.

Along with this transformation, the business model of disseminating artworks has also changed. At this stage, galleries and auction houses try to get involved in the emerging art field of photography, but their traditional industry forms have proved to be a mismatch for the art consumption logic of the image printing stage. To this day, galleries and auction houses are still struggling to sell photographic works effectively. The new chains of bookstores, magazines, newspapers, and the entire publishing system have become a new ecosystem for photographic art.

Phase 3: Mass Media

At the end of the nineteenth century, photography and animation technology combined to create a revolutionary art form – the movie. In 1895, the Grand Café in Paris carried out its first film release. From the 1920s to the 1930s, radio broadcasting technology and television technology moved toward commercial use. The *mass media stage* has flourished ever since.

At this stage, the logic of audience consumption art has undergone fundamental changes: in the manual craft stage, art consumption is "to acquire the property right of original artwork;" in the stage of image printing, it is to buy and appreciate "copies;" in the film stage, although there are still consumers who will collect copies of their favorite movies (previously videotapes, DVDs, and now Blu-ray discs), but for the vast majority of customers, what they buy is not a copy (notably, it is not possible to buy the "original"), but an "experience." What they need is a voucher (movie ticket, which, interestingly, can be a good analogy to today's NFT access to artworks) that can provide them a two-hour enjoyable experience.

Another way to compare the three stages is from the audience perspective. In the manual craft stage, there are very few people who can see the original work. Before the extravagant behavior of international travel became popular, it might be on the order of hundreds to thousands of people. After the popularization of photoengraving and printing, the scale of the audience who could see the reproductions expanded rapidly. In the mass media stage, the audience that a movie can reach is often in the order of millions, tens of millions, or even hundreds of millions.

Phase 4: Digital Art

With the rapid development of computers and chips, in just a few decades, we have witnessed the invention and evolution of a series of technologies: computer graphics, digital display technology, 3D modeling and printing, VR (virtual reality) and AR (augmented reality) technology, artificial intelligence, the internet and social networks, and beyond. These technologies have brought extremely rich possibilities for the creation, consumption, and dissemination of art, for which the NFT is a perfect example.

In another example, video games may become the new frontier of artistic development. In November 2012, MoMA, a state-of-the-art art museum in New York, officially included 14 video games in its collection. It includes names familiar to players of a few years ago: Pac-Man (1980), Tetris (1984), SimCity 2000 (1994). Paola Antonelli, senior director of MoMA's Architecture and Design Department, who oversaw this matter, said:

> Are video games art? They sure are, but they are also design, and a design approach is what we chose for this new foray into this universe. The games are selected as outstanding examples of interaction design – a field that MoMA has already explored and collected extensively, and one of the most important and oft-discussed expressions of contemporary design creativity. Our criteria, therefore, emphasize not only the visual quality and aesthetic experience of each game, but also the many other aspects – from the elegance of the code to the design of the player's behavior – that pertain to interaction design. . ."

The integration of NFT and gaming will be discussed in **Chapter 6**.

NFT and Generative Art

Thanks to NFTs, crypto communities have become an expressive place where users depict themselves in various avatars through their profile pictures. What started as a quirky art project has mushroomed into an entire crypto art movement. CryptoPunks are an internet artifact that inspired the ERC-721 standard that powers digital art and collectibles today. Chances are, you've seen these pixelated punks and bored apes floating in the cyberspace. There are 10,000 unique CryptoPunks, with each boasting randomly generated attributes. These pixelated characters are a mix of guys and girls, rare zombies, apes, and aliens.

The surge in PFPs, the Profile Picture NFTs, also indicates a broader cultural movement. (**See Box: Generative Art.**) From Cryptopunks, PFPs have reached penguins, cats, dogs, monkeys, rats, and more. Trading those JPEG files seem to be steered by the cultural and symbolic value inherent in these creative interactions instead of real-life utility. PFP NFTs allow people to signal status, build community, and encourage artistic expression. Owning one of these NFTs is the virtual admission ticket to internet native social clubs, which continue to grow in size daily.

Generative Art

Just as crypto does, NFT brings with it a host of ecosystem terms. *Generative art* is one such term. According to Agora Digital Art, generative art is the "form of digital art that is generated randomly, whether that is by using autonomous machines or algorithms."

The term *generative art* has been associated with NFTs such as the Bored Apes, Pudgy Penguins, and Wonky Whales, in that these forms of art represent two major themes being explored in generative art: the elements of chance and systems design. As a direct contrast to the more precise discipline and order of classical artists such as Mozart or Monet, generative artists "express themselves through the parameters of randomized systems," according to Coindesk reports.

For example, Joshua Davis is an exemplary generative artist of our time, generating "noisy" art with programming code. Although Davis had established a name for himself in the art scene with his art website Praystation.com (ca. 1995), Davis had not previously thought that his computer-generated art would ever create monetary value, as his art was literally code and not finite, individual pieces as are paintings.

In a 2021 report with CoinDesk, Davis stated regarding NFTs, art, and value: "I thought the next generation maybe would find a way to find value in digital art. I never thought digital art would be embraced as something you could assign provenance, collectability, and scarcity."

The advent of NFTs changed Davis's fate and that of his contemporaries for the better. On the generative NFT platform Art Blocks, artists can upload the algorithms they used to create their art to the platform, where buyers would then have the option of "minting" iterations of the algorithm. When buyers purchase the generative code, they can themselves interact with the code to create their own version of generative art that is entirely unique. According to Shrimpy (a crypto market portal), Art Blocks is hosted on the Ethereum blockchain and is similar to a custom art generator: "[Art Blocks] is like having the artist create custom on-demand artwork for you with the assistance of the blockchain to randomize outputs."

According to CoinDesk reports, Art Blocks has achieved tremendous success for its artists, generating hundreds of millions in sales for artists who previously struggled with generating profits for their work. Similar generative art NFT platforms include 9021, a platform launched in August 2021 carrying 9,021 unique pieces of art with the same pop art theme. According to one37pm, once the first 9,021 pieces all sell out, the second collection of art will be of a new theme, and still be 9,021 pieces in total.

Whereas PFPs are the simplest version of crypto artwork for amateurs, there are also major NFT artists dedicated to the crypto art world. Beeple (aka American artist Michael Winkelmann) – for example, has risen to NFT artist stardom. One of his signature works, "Everydays: The First 5,000 Days," was sold at a Christie's auction in 2021 for $69.34 million. According to an *Indian Express* report, Beeple's artwork attracted great interest not only because it was at the time, the third most expensive artwork sold by a living artist, but ". . . also because it was the first purely digital work offered by a major auction house," in a way legitimizing NFT art as a category in the more traditional art world.

Beeple took a detour before becoming an artist by profession. A computer science major at Purdue, he became a graphic designer, and later began to create and post his own digital art. One such piece is what later became "5,000 days," where Beeple created artwork for 5,000 days, never missing a day in between. Prior to Beeple's debut in the limelight of the art world, he designed artwork for major platforms such as the

Super Bowl and celebrity performance artists, such as Justin Bieber and Childish Gambino.

Fewocious (aka American artist Victor Langlois) is another top NFT artist, at only 18 years old. In the same month of Beeple's 2021 NFT art sale to Christie's, Fewocious reportedly sold $4 million+ worth of NFT artwork on the online art auction platform Nifty Gateway. As recently as March 2020, he managed to sell his first painting for $90 to an art collector who introduced Fewocious to the NFT scene. Fewocious began to sell his art on NFT platform SuperRare, and his art soon reached an average price of 5 ether (around $10,000, depending on ETH fluctuation) per sale. He then began to collaborate with more prominent artists such as platinum-selling musician Two Feet. In only a few years, Fewocious went from a teenage student to being vaulted into NFT art stardom.

For former MLB player Micah Johnson, NFTs served as the springboard. According to the Visa company (a tech partner for Johnson), when Micah retired from professional baseball in 2018, making a living off his art seemed a remote possibility. Micah began learning about crypto and NFTs in 2019 – at the time, a fringe concept. His first NFT sale in 2020, an experiment, marked a major turning point in his career, a way to gain access to an audience outside of the traditional art world. Johnson has built a following and a media business around his crypto-native character "Aku," a young Black boy who dreams of becoming an astronaut. Since Johnson's first debut NFT sale in 2020, his Aku series has spun off into producing physical sculptures of the Aku character, releasing them on platforms such as Nifty Gateway. According to Sporttechie reports, Johnson sold over $2 million of Aku NFTs in 28 hours in 2020 via Nifty Gateway.

Another Johnson-made NFT "Why Not Me" was one of the first two NFTs beamed into the International Space Station and the first NFT character to be optioned for a film. According to NFT curation marketplace Notables.co, the NFT source files for "Why Not Me" were beamed into space, relayed by satellites

Table 5.3   Different Industry Ecosystems at Different Phases of Art History

Four Phases of Art and Tech	Keywords of Art Consumption	Corresponding Industry Ecosystem
Manual Craft	- Appreciation of original, physical objects - Transaction of property rights	- Museums - Art galleries, art fairs, auctions
Image Printing	- Replicas of images - Copyright	- Bookstores, magazines, newspapers, and the publishing system
Mass Media	- Experience instead of ownership	- Film production, distribution, and screening channels - TV business ecosystem
Digital Art	- Immersive experience, interaction, sharing	- NFT marketplaces - Gaming engines and platforms - Social media, incorporating crypto and blockchain tech

to the International Space Station, and then relayed back to the mission control center on earth and minted as authenticated NFTs, traveling over 125,000 miles in space. The metadata containing details of the NFT's flight journey, were built into its own smart contract.

No doubt, an era of generative art is emerging, powered by the blockchain and NFTs. NFTs unlocked an opportunity for everyone to build a community of people interested in supporting his work – in a way that goes way beyond simply liking or sharing in the current Web2. As a new segment of the creator economy, the NFT art market, by some estimates, is the fastest-growing type of small business. Just like museums, galleries, and auction houses in the early phases of art history, NFT marketplaces and more digital infrastructures (See **Table 5.3**) are emerging to accelerate the spread of crypto arts.

Creator Economy: Beyond the Bored Apes

While the prices of individual NFTs remains volatile, creative use cases for NFTs are still emerging and the groundwork is being laid for the long-term utility of NFTs. As the users explore broad application of NFTs, it's worth asking again: what is NFT? Like cryptocurrencies, NFTs are issued on a blockchain, and are used to designate ownership of a certain asset. The process of converting a media file into a non-fungible token is referred to as "minting" an NFT. Each NFT is tied to some unique data, typically a digital content file of some kind (or reference thereto) and governed by a "smart contract." (See **Chapters 2 and 3** for the introduction of "smart contract.")

As such, NFTs can be used to store complex and individual-specific information. Despite the uncertainties of how existing laws and regulations apply to NFTs (more detailed discussion at the end of this chapter), NFTs are an interesting medium for creators of unique contents. The context is that the current version of the internet is not built for content ownership. If you want, you can take any video from YouTube or TikTok, which is hard to detect. But smart contracts change this.

Below is an example of smart contract defining the royalty payment relating to a creative work:

NFT Royalties (EIP-2981: ERC-721 Royalty Standard)

function royaltyInfo(**uint256** _tokenId) **external returns** (**address** receiver, **uint256** amount);

```
/**
 *        @notice Called when royalty is transferred to the receiver. We wrap
emitting
 *                the event as we want the NFT contract itself to contain the
event.
 */
function receivedRoyalties(address _royaltyRecipient, address _buyer,
uint256 _tokenId, address _tokenPaid, uint256 _amount) external;
```

The smart contract can define who is the owner of this content and who are the stakeholders who could share the income from the specific content. For the first time, content on the internet in the form of an NFT can be definitively owned by a specific person independent of a centralized intermediary, and this is unlocking exciting opportunities for digital commerce and engagement. In other words, NFTs are

Seller

List an artwork
from your collection

85% goes to
the seller

Add artwork to
the marketplace

12.5% goes to the
original artist

2.5% goes to
knownOrigin

Listed artworks display on the
KnownOrigin Marketplace

Buy Now

Buyer

Figure 5.2   Example of Sharing Income from a Digital Artwork

turning contents into portable digital assets so that the creator can publish it anywhere on the internet and, more importantly, manage the income streams from the unique content. (See **Figure 5.2.**)

Therefore, while the NFT market started initially with the digital art side, it is going to have broader applications in the creator economy. Beyond the hype of multimillion-dollar digital art sales (such as the bored apes of BAYC), the significance of NFTs may lie in enabling the beginning of a society in the metaverse based on free markets, independent ownership, and social contracts (for goods, services, and ideas). NFTs will be the tool that represents any digital type of assets in virtual worlds going forward, and the applications are tremendous.

For example, NFTs provide significant opportunity for gaming, currently the largest global media, thanks to the metaverse ownership opportunities they introduce. While people spend billions of dollars on digital gaming assets, like buying gaming skins or avatar costumes, the gamers do not necessarily own these assets. NFTs would empower them to play crypto-based games and own assets, earn assets in-game, port them out of

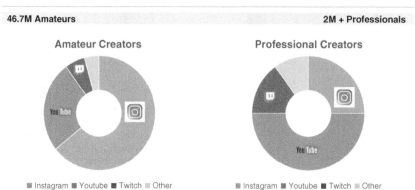

Figure 5.3   Amateur Creators (instead of Professionals) Dominate Metaverse
Source: signalfire.com/blog/creator-economy, end of 2021

the game, and sell the assets elsewhere, such as an open marketplace, as will be illustrated in **Chapter 6**.

This is huge for the creator economy, which is dominated by amateur creators from the public (instead of professionals, see **Figure 5.3**), because the creators can potentially move from a "closed-loop" virtual environment (such as a gaming or social platform) to an "open metaverse." What NFT provides is abstracting digital assets from a walled, closed-loop environment, making it open-loop, and a lot more fluid and free. And we're just touching the surface today on the capabilities that can be built on top of that. Today it is merely trading assets such as gaming items; one day it will give us free movement in the metaverse. As a result, a new social code – powered by ownership and collaboration – is emerging from the ethos of NFT culture. (The open metaverse will be discussed in **Chapter 10**.)

Going Mainstream with Brands and Fashion

By 2022, NFTs have gone mainstream and become an asset class on their own, as compared to crypto tokens, evidenced by NFT Google search interest surpassing that of crypto at the end of 2021. Because of the easy access of NFTs, NFTs and tokenized communities caught the eye of those beyond the stereotypical crypto-savvy individuals. As a result, NFTs have brought the domain expertise of those *outside of the crypto space* to create applications while focusing on *widely relatable use cases.* NFTs especially have vast potential in the content and entertainment world, representing a deeper and more dynamic way to engage fans and potential new revenue streams for individual creators and large corporations.

As discussed earlier in this chapter, the art industry is in the spotlight for NFTs right now; yet other industries are aggressively following suit to create and test NFT products, such as music and sports, video gaming, and ever more consumer brands businesses. In a broader context than crypto assets, NFTs have become a new and innovative way for organizations, brands, artists, and celebrities to engage with their customers and fans. (The gaming industry hosts the perfect environment for the next NFT wave, which will be covered in detail in **Chapter 6**.)

For example, in recent years sports businesses have been challenged to find ways to harness the latest technology and deliver an experience that meets the behaviors of their fans. When COVID-19 restrictions put a swift and definitive pause on live sports, athletes, teams, and stadiums are turning to digital technology to reposition businesses for growth opportunities and to capture the attention of fans. As such, NFTs have emerged as a promising medium for fan engagement. For instance, NFTs can better connect fans to their favorite teams by offering access to exclusive offers, the ability to earn rewards, and even voting rights to team decisions.

In April 2021, the NBA basketball team Golden State Warriors launched an NFT collection, becoming the first team in US professional sports to release their own officially licensed NFTs. Among the collectibles, the Warriors produced a special edition Warriors Golden Ticket NFT combining 75 years of historic franchise moments into a one-of-one ticket stub NFT.

The Warriors believed that there is real value in the legitimacy of something created by an NBA franchise in a marketplace where anyone can create and sell an NFT. "There's a lot more you can do with an NFT than a static sports card," according to a Warriors team executive.

Meanwhile, many major consumer business companies are entering the space and experimenting with NFTs, whether the use case is collecting, betting, trading, gaming, or displaying. This diverse list includes Adidas, Coca-Cola, Dolce & Gabbana, Gucci, Marvel, Mattel, NBA, Nike, and TikTok – to name a few. In retail, NFTs may be used to prove ownership of real-world items, such as designer watches or flash cars, and luxury goods. Further, NFTs can become a new and lucrative revenue stream for consumer brands by offering unique items and engaging top fans. (See **Box: NBA Top Shot and Basketball Card.**)

NBA Top Shot and Basketball Card

NFTs have exploded into sports scenes, and nowhere has that explosion been more visible than on NBA Top Shot, the marketplace where officially licensed NBA digital collectibles that can be bought, displayed, and sold or auctioned by basketball fans and collectors. According to Forbes reports, NBA Top Shot alone was responsible for a third of the $1.5 billion NFT trading volume seen in the first quarter of 2021.

NBA Top Shot is currently the primary project on the Flow blockchain. Flow is another blockchain that is gaining traction for NFTs. While it is smaller than Ethereum, it has started to attract mainstream brands to its infrastructure, as this new blockchain has been built to increase throughput and reduce the challenges of high gas costs through proof-of-staking, a more energy-efficient blockchain transaction validation approach.

NBA Top Shot "moments" celebrate epic game highlights, and include video, action shots, stats, and guaranteed authenticity of ownership. They are essentially digital basketball cards, but instead of static images, these NFTs contain video highlight moments from NBA games. If you compare a basketball card to a Top Shot moment, you start to realize why people like NFTs.

For basketball cards, you have to send them somewhere to get graded, which could easily take six months to a year, and must store them somewhere. Then you have to figure out where to sell them, if you still remember where they are stored. You don't know how many are created, you don't know what the card has sold for before, who's owned it, and so on. Then compare that to the NBA Top Shot moment NFT. You know everyone who has owned it and what price it sold for; storing it obviously is not an issue at all. You don't have to worry about getting anything graded. That is full transparency vs. complete opacity.

These are not simply memes or one-offs. For example, NFT-controlled access could span a range of use cases, including VIP access to real-life conventions and festivals as well as those occurring within the metaverse. They could also be used for airdropping branded merchandise or allowing special access to fan-only content, potentially opening an entirely new avenue for fan engagement.

The sports brand Adidas is a great example, where NFT buyers will get access to "digital and physical" Adidas products and experiences. At first, the physical goods will include the tracksuit worn by Indigo, a hoodie with a blockchain address on it, and an orange beanie. The product is cobranded with a trio of collaborators: Bored Ape Yacht Club, the highly sought-after NFT collection; Punks Comics, which recently featured Indigo on an issue's cover; and GMoney, a pseudonymous crypto enthusiast who has been consulting with Adidas on how to enter the NFT space in a way that feels authentic. The beanie is supposed to be the one worn by GMoney's CryptoPunks avatar.

Another example is blanksoles, an online fashion platform that crosses between both the NFT world and the physical fashion realm. blanksoles, according to their official website, is the platform for sneakerheads to launch both physical and digital collections and will be the future of fashion in the metaverse. Their central value proposition is that by owning a blanksoles (or designsoles) NFT, the buyer also receives a physical designer shoe, and membership to blanksoles' collective of iconic musicians, artists, athletes, and other notable individuals.

According to NFT marketplace Magic Eden, members of the blanksoles community will receive a blank white shoe NFT and physical shoe, which acts as the template for future customizations for additional drops. blanksoles offers two types of NFTs, blanksoles and designsoles, where blanksoles give the holder access to mint a pair of designsoles for free, but the user must burn the blanksoles to do so, and designsoles are designer shoe NFTs that will be created by the world's most renowned artists and collective members. designsoles NFTs will be accompanied by a pair of physical shoes.

The blanksoles brand strategy is to be the leader of the future of fashion in the metaverse era, by making use of both the growth momentum behind NFTs and the metaverse, and the more traditional model of designer footwear. The genesis limited-edition blanksoles mint occurred in December 2021 for its early community members, and the first designsoles mint occurred January 2022.

Going forward, NFT will become a new tool for consumer businesses, at a time when the line between physical and virtual experiences is blurring – spaces and exhibits with a mix of physical and digital content are growing, and consumers perceive a brand's digital presence equally as important to its in-store presence. Furthermore, consumers are increasingly interacting with companies and brands in fully digital environments, for example in video gaming through "skins" and other in-game items, where NFT can play an important role. Every brand may need an NFT strategy, or risk becoming less relevant.

Challenges to the NFT Metaverse

In summary, NFTs represent one of the most exciting, fast-growing areas of the cryptocurrency world. When Facebook changed its name in October to Meta, the metaverse garnered so much attention that 2021 was dubbed the "Metaverse Year." But we should also see 2021 as "NFTs: Year One" – at the beginning of 2021, only a niche group of crypto enthusiasts knew what NFTs were; but by the end of the year, they were especially popular with retail investors, and the increasing participation of major brands and corporations brought NFTs to the mainstream.

Today we already have a nascent version of the Metaverse existing with digital goods like NFTs representing popular art and digital memorabilia. The NFT collections are developing complete ecosystems and users are getting more engaged with them. They were more than just digital jpeg photos, too; verticals like gaming, music, and sports merchandise grew increasingly eager to integrate NFTs into their ecosystems. Growing entry of major players including The Home Depot, Microsoft, Starbucks, Tesla, and Whole Foods remain a key driver of growth.

Still, the NFT market must overcome three major challenges to succeed going forward, from speculation to mainstream adoption:

First, minting / gas fees for NFTs currently serve as a barrier to entry, preventing NFTs' mainstream adoption. Based on public data, Ethereum is the dominant smart contract platform and the blockchain of choice for NFT issuance and transactions. But while the massive surge in demand in 2021 was a huge net positive for the industry, the increased network activities from NFT mania pushed the Ethereum network to its breaking point, rendering it unusable for many retail users due to high gas fees and scalability problems, creating demand for "layer 2" scaling solutions to lower Ethereum's transaction fees while maintaining security.

As fees continue to rise on Ethereum, however, several alternative protocols offer creators lower fee alternatives (e.g., Tezos, Near). Going forward, the competition from other layer-1 blockchains and layer-2 scaling technologies is likely to increase. Furthermore, interoperable blockchain infrastructure will also emerge (such as Polkadot, which works on para-chain solutions) to support cross-platform usability of NFTs, where NFTs can be bought and transferred across platforms (e.g., social media and video gaming platforms). The competition among the new blockchains should lead to more user-friendly infrastructures, lowering the barrier of entry for average NFT players.

Second, critical infrastructures are still missing from the current NFT market, and for the existing ones, they are very "centralized." Despite the remarkable rise in value of the NFT market recently (reaching over $40 billion in sales in 2021), the channels for users to realize the special NFT value remain incredibly limited. Most NFT transactions simply involves trading NFTs on exchanges like OpenSea and profiting from price movements. To fully realize the value of NFTs as a financial asset class, it's critical to develop more applications in DeFi that create more ways to compose these assets in holders' investment portfolios.

Furthermore, there is some exhibition infrastructure missing to fully realize the vision of NFT creator economy. For instance, artists and marketplaces need to better empower conspicuous ownership, because public exhibition is an important source of value for NFT owners, just like the collectors of classic arts. Owners need platforms and tools to display the artworks they have purchased to others that is not just a link to an OpenSea page.

Also, the decentralization of the algorithm behind NFT and the commercial operation of the NFT are completely different and should not be confused. The best example is OpenSea, which is the first and largest marketplace, similar to eBay for NFTs, where users can buy and sell pieces from almost any NFT collection out there. According to DefiLlama NFT data, by the end of February 2022, OpenSea had contributed to approximately $20 billion NFT transaction, equivalent to about 97% of total volume of all marketplaces.

(DefiKingdoms, MagicEden, and ImmutableX are the three marketplaces immediately following OpenSea, but each of their trading volume was only slightly more than $100 million.)

The OpenSea marketplace is a perfect example that the functions of traditional art intermediaries (such as galleries and auction houses) for the discovery, marketing, and transaction of artists and artworks are still important in the digital age. In the era of digital art, the top NFT trading platforms will surely become the new "central force" in the emerging Metaverse. This raises important governance questions about the "open metaverse" concept, and in **Chapter 10**, we will discuss new decentralized autonomous organizations (DAOs) that are challenging centralized infrastructures like OpenSea.

Third, the biggest obstacle for NFTs is the legal uncertainty around such "digital asset." In fact, many consumers may have no idea what it is they're buying. Are NFTs virtual currencies? Or, are NFTs a certificate for virtual currencies? And more importantly, are NFTs securities? (In May 2021 the NFT players in the US sent a rulemaking petition to the SEC to seek an SEC concept release to resolve regulatory uncertainty.) These are the questions that no major digital economy's legislature has ever answered. The unregulated nature of crypto assets and NFTs can drive potential players and investors away.

As described earlier in this chapter, most nonfungible tokens are a metadata file that has been encoded using a work that may or may not be subject to copyright protection, or it could even be a work in the public domain. Anything that can be digitized can be turned into an NFT; the original work is only needed in the first step of the process to create the unique combination of the tokenID and the contract address. So, in principle, NFTs have very little to do with copyright. NFT ownership is not intuitive.

The key issue here is the often-widespread confusion surrounding the rights that buyers acquire when they purchase an NFT. Some buyers think they have acquired the underlying work of art, as well as all its accompanying rights. However, in reality, they are simply buying the metadata associated with the work, instead of the work itself. Some of the confusion may be caused by the amount of money spent on the NFTs. When generative art works can be sold for over US$ 1 million, it is easy to assume that the purchaser has acquired more than a short string of numbers and letters of dubious artistic value. (See **Box: NFT Skybound.**)

Furthermore, when the NFTs of existing artwork have had multimillion-dollar price tags, less tech-savvy people often assume that it is the work itself that has been sold, which is not the case. Understandably, it is difficult to comprehend that buyers of NFTs are spending such large sums of money on what amounts to a metadata file without any linkage to the copyright of such arts, but that's exactly what most NFTs are. Average users therefore need to be clear that the main reasons to buy an NFT are the same for any "collectibles": potential investment return from owning an illiquid asset and the pleasure of having something unique from an admired artist, brand, sports team, or whatever. Unless the terms allow it, buyers will only have a limited ability to share the creative work on public platforms or to reproduce it and make it available for others.

NFT Skybound

In China, application and commercial deployment of NFTs have found their way to the payment giant, Alipay. June 2021, Alipay, the fintech arm of China's internet giant Alibaba, and Dunhuang Art Museum collaborated to launch 8,000 limited-edition NFTs, "Feitian (Skybound)" and "Luwang (Deer King)", that act as decorative skins for the Alipay app and are commemorative of Dunhuang's ancient cave art. Tencent, the internet and gaming giant of China, followed suit two months later with its NFT app Magic Core, or *Huanhe* in Chinese.

The first batch of Alipay NFT art consisted of images of flying apsaras and sacred deer from the Caves of the Thousand Buddhas, or Mogao Cave, in the Chinese city of Dunhuang. Buyers were able to purchase the artworks using 10 Alipay points plus CNY9.90 (US$1.5), essentially using fiat money instead of cryptocurrency. The NFTs were sold out within 24 hours after Alipay made them available to the public, even though as stated by Alipay, the art copyright belongs to the original creator rather than the purchasers, and the NFTs cannot be traded or used for any other commercial purpose.

Subsequently, many users began to sell Alipay Skybound NFT on the Xianyu website, an e-commerce platform under Alibaba for used goods. Millions of second-hand transactions have appeared on Xianyu, and the NFT price NFT was fired up to CNY 1.5 million (approximately a quarter of US$ 1 million). On the second day, the Xianyu platform quickly removed Alipay Skybound NFTs to avoid regulatory issues. (China has a strict prohibition on crypto trading, which will be discussed in detail in **Chapter 9**.)

There is a core difference between how NFTs are sold inside and outside China. In the West, digital collectibles like NFTs are minted on pubic open blockchains; by contract, in the case of Alipay Skybound NFT, Alipay's private blockchain AntChain (also known as "private consortium blockchains") provided the blockchain technology services for the data storage and unique identification of the NFT artworks. Unlike Ethereum or the bitcoin blockchain, the blockchains in China are managed by a select number of China's tech giants, which are not open for the public to participate in and authenticate the data.

Given the sensitivity around crypto, Chinese tech giants have been cautious. Alibaba and Tencent started using the term *digital collectibles* (instead of NFTs) in October 2021. NFT platforms in the country have also banned second-hand selling. For example, the AntChain allows NFT owners to transfer their assets to another party 180 days after the purchase. The second owner can transfer the asset after two years. However, selling the NFTs for money is still prohibited, and ownership is limited to mainland residents who are at least 14 years old.

According to recent Chinese news reports, some Chinese users have started to think of giving up on buying "digital collectibles." After all, what is the value of a few digital photos that one cannot sell?

Now, if we go back to the beginning of this chapter and look at this NFT phenomenon in the context of co-evolution of art and tech, we may find the NFT economy a lot like the traditional patronage model of art funding. But the NFT could be *much better.*

Historically, patronage involved wealthy patrons directly supporting individual artists, which is similar to an artist receiving one major payment from a wealthy NFT investor. The downside of patronage funding is that the artists were hostage to patrons, a narrow elite few of the society. Furthermore, although there is a long tradition in the art world of owning art and lending it to a museum – appropriately acknowledged – for the public to enjoy, a lot of patron-funded art is held in private storage facilities as private collections, where they are seen rarely if at all.

Today, NFTs offer the benefits of patronage without the well-understood costs. Wealthy individuals can continue to support art and enjoy the experience of ownership. Rather than relying on a small community of the rich in Florence like Renaissance artists, digital artists can easily reach a global supply of patrons. The ownership of art can be demonstrated (cryptographically proven, powered by the blockchain) to all who the patrons seeks to impress with their impeccable taste. But at the same time the art itself remains free for the public to enjoy. This looks like the old patronage economy for creators to financialize their work, but at the same time leave them open access for anyone to view, watch, or listen to. (Phase 1 in **Figure 5.4**.)

This is huge for the creator economy, where NFT buyers and sellers can determine market-clearing prices on blockchains instead of data aggregation platforms, creating new forms of asset monetization. For example, in the eyes of musicians, the current music industry model is broken. The problem started

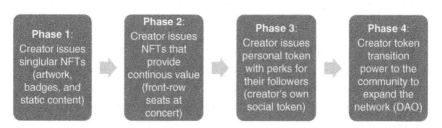

Figure 5.4   **NFT example – The Creator Evolution Timeline**

in the days of Napster, before Apple came along and decided an MP3 is worth 99 cents. But why should Apple decide how much music is worth? Furthermore, streaming is unsustainable for most musicians because revenues tend to be divvied up pro rata. According to *Rolling Stone,* 90 percent of streams are going to the top 1 percent of artists on platforms like Spotify. In a pro rata model, that 1 percent ends up with 90 percent of the money these platforms have decided to set aside for artists.

Now the audio NFTs are starting to solve these problems. The artists can price their music however they choose, as the NFTs provide musicians a direct link to the fans, departing from the platforms (and labels) that are beyond their control. For example, Kings of Leon recently announced plans to release an album as an NFT, and the artists can bundle the NFTs with real-world perks. As *Rolling Stone* reported, the most exclusive NFTs will include a "golden ticket" so token holders are guaranteed "four front-row seats to any Kings of Leon concert during each tour for life." (Phase 2 in **Figure 5.4.**)

Furthermore, the NFT itself is evolving, too. At the beginning of 2022, collectibles and digital art account for more than 75 percent of NFT sales on Ethereum. By contrast, NFT sales in virtual worlds like The Sandbox and games like Axie Infinity have accounted for less than 25 percent of cumulative sales on Ethereum. However, as will be discussed in **Chapter 6**, with the booming of video gaming market, NFT demand for blockchain-based games and virtual worlds is skyrocketing, especially as NFT collectibles begin to exhibit more utility in various games. (Phase 3 and 4 in **Figure 5.4.**)

To conclude, NFTs in 2022 are expanding from digital art into numerous applications. As the public has gotten comfortable with NFTs, conversation has turned to further virtualization opportunities. In the Metaverse, everything can and will have its NFT, and NFTs offer a liquid marketplace in which consumers can invest in different digital assets and engage in peer-to-peer transactions.

Eventually, NFTs will blur the line between consumption (such as gaming entertainment) and investment (such as play-to-earn). The combination of NFT and metaverse could open up the possibilities of full-functioning economies, and hence, attract a larger user base, which will be illustrated by the gaming industry in the following chapter.

Blockchain Gaming in Metaverse

- From Gaming into 3D Interactive Metaverse
- Tech Convergence, Media Convergence
- Epic Games and Fortnite UGC
- Roblox Human Co-Experience
- P2E Gaming – GameFi with NFT
- Blockchain Gaming: Gaming First? Crypto First?
- Gaming, the Foundation of Metaverse

From Gaming into 3D Interactive Metaverse

Human beings have been playing games for thousands of years. From as far back as 3100 BCE, the ancient Egyptians were playing a board game called Senet, and China is the birthplace of *weiqi* (*Go* chess), which was played as early as the Zhou dynasty (1046–256 BCE), involving more possible board configurations in *Go* than there are atoms in the visible universe. It seems that people naturally crave a challenge of wits and physical skill as "interactive entertainment," and society has developed elaborate social constructs to cater to these basic human desires.

As we entered 2022, interactive entertainment – mostly represented by video gaming on mobile smartphones – layers centuries of technological advancements on top of the core human interactions these ancient games required. No matter where you are on earth, you can enjoy a challenge of wit and skill at the speed of light, thanks to the global network of mobile internet. Participants can play against another person, a machine, or even themselves, and the entertainment is interactive, instant, and engaging.

Before gaming, online entertainment content was consumed in the past tense. Photos, tweets, videos, and movies are all a part of our lives that are captured first and then shared and consumed by others afterward. Gaming, on the other hand, is inherently a real-time social experience, whether it is with friends playing together online, or streamers broadcasting to their audience live, esports teams competing with their peers. When "Metaverse" denotes a shared, virtual space that is persistently online and active, mirroring our real physical world in the digital realm, the term sounds quite like gaming.

The future of entertainment is interactive, real-time, and 3D, as promised by the Metaverse. 2D, asynchronous, noninteractive, static content is losing the attention of the young digital native generation. The world is changing. Leaps forward in computing power and bandwidth are enabling an explosion in interactive, real-time 3D content-led by games and is now spreading rapidly into other industries.

Today, games are a massive global industry and expand across demographics. With the rise of the mobile internet, gaming ascended from a niche hobby to a global phenomenon (see **Figure 6.1**) beyond narrow, outdated stereotypes ("crazy kids and niche market"). The video game market is substantial – an industry that is now larger than movies and music combined around the world. Gaming is now the world's most valuable media category, most watched sport, and most active social meeting space of choice for an entire generation – both the Gen Z and surprisingly, the adults.

According to IDC 2021 data, the $203 billion gaming industry is the fastest-growing part of the global media and entertainment industry and has grown 19 percent in 2020. IDC also

Gamers Now Approaching 3 Billion Global Players

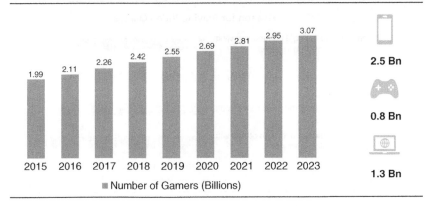

Figure 6.1   1.3 Billion Gamers, Globally
Source: Newzoo

reported that mobile gaming represents approximately half of the gaming category and grew 22 percent year-over-year, faster than the console gaming and PC gaming sectors. Globally, nearly 40 percent of gamers are over the age of 35, which may shock non-gamers who tend to think gaming is all about "crazy kids." This user base is also highly engaged. As many as 60 percent play daily, with the average player playing for over six hours a week, according to the ESA and Limelight Networks, respectively.

Therefore, gaming is the future of social network. Gaming is where people connect with their friends to have great social experiences offline, like going to a museum, hanging out at a park, or working out together, which is a richer social experience than sitting on a couch and talking. That would be a key ingredient for the future metaverse, and it has been proven out in dozens of different genres of games. For example, popular games like "League of Legends" has more than 100 million active players monthly and it has its own world championship tournament, hosted both online and at physical sports stadiums, just like soccer's World Cup.

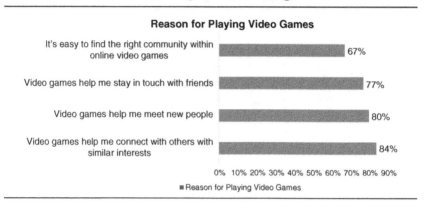

Figure 6.2   Gaming Is the Next Social Network

Source: Adapted from Accenture, data complied by Goldman Sachs Investment Research, 2021

Similar to previous waves of digital technologies, including online search, social, e-commerce, mobile, and short videos, gaming will fundamentally change how people interact with each other and the internet overall (see **Figure 6.2**). For the young generation, navigating a 3D environment and hanging out in a virtual world with friends is all commonplace. In the Metaverse, social experiences and functionality will take place in the virtual gaming world, such as:

- Messaging
- Live broadcasting
- Celebrating life events (weddings, birthdays, etc.)
- Social (spending time with friends or hosting parties)
- Forming relationships (making new connections based on shared interests)
- Dating

As such, gaming is where culture will be created and set, evolving into a deeply social behavior across online and offline worlds. In this chapter, we examine how the gaming world, with the new addition of blockchain technologies, has already shown some key elements as to how the Metaverse might evolve.

Tech Convergence, Media Convergence

Over the past decades, games have become a catalyst for technological innovation. Since breakthrough advances in microelectronics allowed the creation of the first video games in the 1970s, play and technology have formed a symbiotic relationship. While better technology enabled increasingly immersive gaming experiences, those games – in return – have accelerated an era of exponential technological progress.

For example, artificial intelligence (AI) has evolved with machines competing with human players in various chess games. Decades ago (1997), IBM's Deep Blue beat chess grand master Garry Kasparov. However, *weiqi* (*Go* chess), mentioned at the start of this chapter, is a much more difficult game for computers to master. Years later, when the AI-enabled computer Go program called AlphaGo, designed by the DeepMind Lab of the US internet giant Google, beat the best human player in 2017, it was viewed as a huge milestone for AI research. (**See Box: Match Impossible.**)

Match Impossible

In May 2017, Chinese media was abuzz with reports about a historical match of the Go chess game (*weiqi*). It was a best-of-three match between Chinese player Ke Jie, the world's No.1 ranked player and world champion, and the AI (artificial intelligence)-enabled computer Go program called AlphaGo, designed by the DeepMind Lab of the US internet giant Google.

China is the birthplace of *weiqi*, an ancient board game played on a 19 × 19 grid. In *Go*, two players place black or white stones on the grid, each seeking to seal off the most territory. Historical records show it was played as early as the Zhou dynasty (1046 BC–256 BCE). The match took place in Wuzhen, Zhejiang province, where there is a canal more than 1,300 years old – a fitting venue for a game that dates back thousands of years. Wuzhen also hosted China's annual World Internet Conference, creating a parallel link to the digital power of AlphaGo.

For many, the Wuzhen showdown was ripe with suspense and symbolism. Human versus machine; tradition versus modern; intuition versus algorithm; East versus West. Who would prevail?

In contrast to the long history of Go within Chinese culture, AlphaGo was only three years old at the 2018 match. Go is seen as an extremely difficult game for computers to master because there are more possible board configurations in Go than there are atoms in the visible universe. Furthermore, human players believe that winning multiple battles across the board relies heavily on intuition and strategic thinking. The idea that a software algorithm could memorize all combinations of board pieces, assess the situation by calculating all possible moves, and select a winning strategy seemed out of the realm of possibilities.

As such, the Go game has been a benchmark for measuring the human mind against artificial intelligence after IBM's Deep Blue beat chess grand master Garry Kasparov in 1997. For many years, there was little progress. More recently, the AlphaGo program developed by Google's DeepMind managed to analyze the game in a different way. AlphaGo used two sets of "deep neural networks" containing millions of connections similar to neurons in the brain – one that selects its next move while the other evaluates the decision.

The Google programmers provided AlphaGo a database of 30 million board positions drawn from 160,000 real-life games to analyze, and the program was also partly self-taught, having played millions of games against itself following initial programming (machine learning), all the while learning and improving. AlphaGo's success was considered the most significant yet for AI due to the complexity of Go game.

In the end, a tearful Ke Jie became the hallmark image of this historical match. After losing 0–3 to AlphaGo, Ke Jie took off his glasses and wiped his eyes, his crying filling up the room where he had fought and lost. Meanwhile, the DeepMind Lab team announced that AlphaGo would retire from competing against human players. Instead, the team would largely shift toward using AI to solve problems in health, energy, and other fields.

Take 3D contents for another example. For almost a hundred years, photos and video content have largely been created by the same means – capturing three-dimensional images through a 2D lens and projecting them onto a 2D surface. Even with the latest digital cameras, the basic processes of content creation remain around building 2D, asynchronous, non-interactive, static content. Driven by gaming needs, companies like Unity and Epic Games have developed 3D gaming engines for developers to create 3D contents, and such technologies are then picked up by education and real estate construction industries.

Gaming now finds itself at the convergence point of advanced tech. Partly driven by the Covid-19 pandemic that forced people indoors and glued them to their screens, the gaming expanded exponentially during the past two years, empowered by the latest technologies such as mobile internet and smartphone, 5G connectivity, cloud computing, and of course, blockchain:

- **Mobile internet and smartphones.** Gaming is rapidly moving into the mainstream as a wide variety of games become available on mobile devices. Even some graphics-intensive games can now be played on the latest smartphones, allowing users to enjoy the experience with or without an expensive gaming console or PC. The young generation users leapfrog into mobile, and their uptake of interactive entertainment has grown especially on smartphones
- **5G connectivity.** The 5G network promises fast streaming and low latency. A large percentage of consumers will upgrade to 5G mobile devices and higher bandwidth speeds. Video game companies like Epic Games, Unity, or Niantic Labs are building the simulation software and concurrency infrastructure that will allow billions to co-experience synthetic reality. Processing and networking advances are enabling more sophisticated, immersive, and social environments.
- **Cloud gaming.** Cloud computing has removed the limitations on processing power and storage to support content. These factors are driving significant increases in available compute power and are enabling the spread of immersive and interactive content. Cloud gaming is driving rapid growth of the user population.

Blockchain and digital assets represent the cutting-edge infrastructure level revolution within gaming. The crypto and blockchain technologies are set to disrupt the games and digital entertainment space in a profound way, as gaming content creation and in-game digital assets will broadly move onto the blockchain. Such a new digital ownership paradigm is unlocking new experiences, user acquisition strategies, and business models, as illustrated by play-to-earn (P2E) models. Before the discussion of blockchain gaming, let's look at the top two gaming apps by user time spent, Fortnite (Epic Games) and Roblox, who are morphing into Metaverse platforms and are beginning to show us what it is like to be in the Metaverse.

Epic Games and Fortnite

Epic Games, based in Cary, NC, was founded in 1991 by Tim Sweeney, the company's chief executive. Epic's most recent funding round came in 2021, where it raised $1 billion in funding, valuing the company at $28.7 billion. Sony, the creator of the PlayStation game console, invested $200 million, and more interestingly, blue-chip institutional investors including Appaloosa Management, Baillie Gifford, and Fidelity Management were also among the investors. The enormous size of the Epic financing round is a clear indicator that the capital market believes in the convergence of gaming and Metaverse. (Tencent, the Chinese internet giant, owns a 40 percent stake in the company.)

Like Unity, a firm founded in Denmark in 2004, Epic Games developed a successful game engine named Unreal Engine, a platform gaming developers could use to create video games. The Unreal Engine is in direct competition with Unity's game engine for being the most popular software to power games. At the same time, both are promoting their products as general-purpose simulation software that they hope will become a common language in which 3D worlds are built, in the same way html underpins websites.

Epic Games' breakthrough came in 2017, when it released Fortnite. The animated, battle royale-style title *Fortnite Battle Royale* is a free-to-play game in which up to 100 players fight to be the last person standing. It has become one of the most popular video games, and has further spawned a new generation of live streaming. It made gamers who broadcast their play of Fortnite, like Tyler Blevins (known as Ninja) into a new breed of online influencers and wealthy celebrities, similar to the popular short video influencers on TikTok.

Fortnite has been an early adopter of interactive technologies, with crossovers from the real world. For example, it has organized a set of interactive in-game performances by mainstream artists, drawing tens of millions of viewers. (**See Box: Rapper inside Fortnite Game.**) Fortnite has been flagged as an early example of what a metaverse might look like. "We don't see 'Fortnite' as *the* Metaverse," says an executive of Epic Games, "but as a beautiful corner of the Metaverse."

Rapper inside Fortnite Game

One celebrity that has made a high-profile appearance in the Metaverse world is the rapper Travis Scott, who made an in-game concert performance titled "Astronomical" in Fortnite. Mostly known for his larger-than-life live concerts, he used Fortnite gaming platform to expand that persona into a surreal experience and a visual marvel.

In the 2020 Fortnite concert, Travis Scott was depicted by a 200-foot-tall version of himself. Attendees of the concert could choose to bounce, float, swim, or fly around the entire Fortnite map (which was at one point submerged in water), while Travis performed songs walking around the map in his giant virtual self. Over the course of the 10-minute concert, Scott also led millions of virtual Fortnite players and visitors underwater and into outer space. The concert show made use of both the immersive element of the Fortnite gaming experience and Travis Scott's real-life rapper persona to create a memorable, metaverse experience for Fortnite players.

How is the Fortnite concert on a gaming platform different from a conventional virtual concert? For one, the Fortnite concert offered more interactive elements for concert attendees than would a conventional online concert. Attendees of the Travis Scott performance move, swim, and fly around the

Fortnite map and "party" with their own virtual avatars. Also, the 10-minute Fortnite concert was much shorter than an average concert.

Another key difference is that the concert is in-game, whereas a traditional on-stage concert is a one-time event. In the metaverse of the Fortnite game, perhaps after attending the Travis Scott concert, users will become even more immersed in playing Fortnite. After a one-time concert ends, it is up to the management team behind the performer to try and capture fan loyalty, but Fortnite would have already captured those loyal users.

Increasingly, Epic Games/ Fortnite considers itself a platform that gives users tools to create and monetize their own games, as well as travel between thousands of worlds within a so-called *Multiverse* (a term similar to Metaverse). For example, the Creative Mode of Fortnite (known as *Fortnite Creative*) gives complete freedom to players to design Fortnite games that can be published and shared with friends online. *Fortnite Creative* offers a wide range of tools for users to design user-generated content (UGC) in Fortnite, and the users can implement their own rules on their own UGC "personal islands." They can also play countless community-made games with friends by entering existing "islands."

In the past, UGC was typically viewed as poor in quality. However, with the development of affordable and easy-to-use hardware, such as digital camcorders and mobile devices with high-resolution video cameras, as well as advances in software technology such as desktop editing software, the barrier for producing quality content is rapidly decreasing. Today's UGC can also be viewed as professional (quality) user-generated content, or PUGC, which combines the content breadth offered by user-generated content (UGC) and the quality offered by professional-generated content (PGC).

From Epic Games/Fortnite, we can see that the popular culture is shifting – users are content creators themselves, which makes them the unequivocal center of social entertainment platforms (vis-a-vie entertainment companies that supply big-budget productions). Being a "creator" of UGC is an increasingly desired career, especially across younger genera-tions. This boom in the supply of high-quality creators into the space brings on more success to the ecosystem, and in turn, an increase in the supply of new media we consume from indi-viduals instead of large companies. In other words, the UGC trend is breaking the model of the traditional gaming industry, which is further illustrated by the Roblox case in the follow-ing section.

Roblox Human Co-Experience

Roblox is like a YouTube for games. Roblox is a massive user-generated social gaming platform that allows kids and teens to create custom games and play them with their friends. It is both a game and a platform. The company's eponymous "Roblox" game was the biggest mobile game of 2020 in the US in terms of revenue, according to Sensor Tower's data. It surpassed "Candy Crush," which had been the top game for the past three years. But what makes Roblox different from other online game platforms is that it lets users create their own games. "Every experience is built by our community," says Roblox.

Just like the emergence of generative art with NFT develop-ment seen in Chapter 5, expanding UGC among normal gam-ers is a powerful trend at Roblox. The platform allows users to operate a virtual avatar on the platform to create and play games built by other users. Upon signing up for Roblox, users personalize their avatars by selecting body types, clothes, and gear. Users are then free to immerse themselves in the mil-lions of developer-built experiences. The platform produces 20 million games a year and has a large economy with creators developing and selling accessories, gears, and items for cus-tomizing avatars.

CEO of Roblox, David Baszucki, has often referred to Roblox as a "human co-experience," a term indicating that the Metaverse is bigger than gaming, which predicates on the following fundamentals: identity, social, immersive, low friction, variety, anywhere, economy, and civility. Roblox considers the "human co-experience" in 3D digital worlds to be the new form of social interaction that its founders envisioned back in 2004. In the "human co-experience" space, users can do things together, such as work, learn, play, shop, and experience entertainment – way more than gaming itself.

The platform is now working on turning the platform into a metaverse that would transform the gaming space into a virtual world seen through VR headsets. For example, together with Gucci, Roblox set up the Gucci Garden in May 2021, an immersive experience to celebrate Gucci's 100th anniversary, and users could buy and wear limited-edition Gucci virtual items for their avatars across the Roblox platform. One notable Gucci Dionysus bag with a bee was sold for US$6 but afterward resold for over US$4,000. Roblox also has leveraged its platform to host several virtual concerts throughout the pandemic, including Paris Hilton's New Year's Eve party. (**See Box: "Paris World" in Roblox**.)

"Paris World" in Roblox

More celebrities began to jump on the metaverse bandwagon since 2021. In December 2021, socialite Paris Hilton hosted a metaverse New Year's Eve party on her virtual island "Paris World" in Roblox. Hilton previously stepped into the NFT arena, working with designer Blake Kathryn to design and sell three unique art NFTs. Sold in an online auction, the most expensive piece sold for more than $1.1 million.

The New Year's Eve party marked Hilton's debut into the metaverse world. At the event, a virtual Hilton avatar played an electronic set to entertain her fans and other visitors to her virtual island. The island invited the visitor to a metaverse exploration of Paris Hilton's life, including a replica of her Beverly Hills estate and dog mansion. The island could be explored in a luxury sports car or yacht.

All these interactive elements combined to create a far more immersive experience than would a New Year's Eve party on virtual conference software such as Zoom. Players could even stroll through the same neon carnival scene location where she married her husband, Carter Ruem, in 2021. Paris World tourists could buy virtual clothing or special rides (e.g., on a jet-ski) on the island as well.

On Zoom, the performer can also interact with party attendees, but the experience will be by far less metaverse-like than a virtual Paris World could offer, complete with scenic wedding backdrops and jet-skis. Paris Hilton's metaverse island party delivers both Paris Hilton and what it would be like to live in her world as a one-package experience, whereas over Zoom, it would be a relatively simple delivery of what it's like to have Paris Hilton in (virtual) attendance.

Paris Hilton envisions Paris World to be an extension of her real-life world, now an experience available to more people across the globe due to the digital metaverse option. Reuters quoted Hilton: "For me, the metaverse is somewhere that you can do everything you can do in real life in the digital world. Not everybody gets to experience that, so that's what we've been working together on over the past year – giving them all my inspirations of what I want in that world."

Constantly trying to improve the "human co-experience" at its platform, Roblox focuses on developing behind-the-scenes tech capabilities for an immersive 3D environment. The Roblox platform provides server space and digital infrastructure to support shared experiences for an average of 30 million daily active users, ranging from how these experiences are built by an engaged community of developers to how they are enjoyed and safely accessed by users across the globe. It is composed of three elements:

- **Roblox Client:** The application that allows users to explore 3D digital worlds.
- **Roblox Studio:** The toolset that allows developers and creators to build, publish, and operate 3D experiences and other content accessed with the Roblox Client.
- **Roblox Cloud:** The services and infrastructure that power the human co-experience platform.

In the Roblox context, experiences refer to the various titles that can be enjoyed by Roblox's users on the platform. Users who create experiences are called developers and those who create avatar items are called creators. Developers can also build and sell custom tools and 3D models to help other developers create experiences. In the same way, firms such as TikTok provide tools that allow average users to look like professional stars in self-made short videos, Roblox comes with a set of easy-to-use programs that let rookies build and monetize their own 3D games and experiences.

Collectively, Roblox's developers and creators contribute to their platform in three ways: by building experiences for users to enjoy, by building avatar items for users to acquire and express themselves with, and by building tools and 3D models for other developers and creators to utilize. The types of content on the company's platform have broadened over time, and developers of those virtual items receive around 30 percent of the proceeds generated from a game, such as the sale of virtual outfits and avatars.

Roblox's business model is largely centered on users' purchases of virtual currency that allows them to acquire in-game perks or virtual items for their avatars. It has its own currency, called Robux, which is paid for with real cash. Users can spend it in what is, in effect, an app store that sells the powerups or cosmetic items like shirts, hats, or pairs of angel wings, which avatars need to stand out.

According to a survey conducted by Newzoo, consumers view the ability to choose their avatar's physical appearance as a key feature in terms of driving overall enjoyment within the Metaverse, followed by free content funded by advertisers and sponsors, and the ability to create content for other players. Despite representing the minority of revenue, avatars are a key element of the Roblox experience as 20 percent of users change their avatars daily (according to the company's data) – the more personalized a gamer's avatar is, the more engaged they are, the more invested they are in the platform, and the more time they spend.

Meanwhile, advertising revenue from major consumer brands is also significant. To create its in-game brand activations,

Roblox often connects brand partners with developers working within the platform, which creates more channels of income for the developer and creator community. As the Roblox platform has scaled, the monetized developers and creators have enjoyed meaningful earnings expansion over time, which drives a growing incentive for such developers and creators to continue to build high-quality content.

This has led to a positive feedback loop at the platform. As developers and creators build increasingly high-quality content, more users are attracted to the Roblox platform. The more users on the platform, the higher the engagement and the more attractive Roblox becomes to developers and creators. With more users, more Robux (the Roblox currency) are spent on the platform, incentivizing developers and creators to design increasingly engaging content and encouraging new developers and creators to start building on the platform.

This robust ecosystem has attracted consumer brands; for example, Nike has also teamed up with Roblox to establish a virtual world called Nikeland. Recently, something revolutionary has occurred: some metaverse-minded brands have begun to skip the middleman of Roblox, partnering directly with the developers to bring their products into virtual spaces. This is leading to a vibrant and expanding creator economy on Roblox, defining new relationships and creating new opportunities among brands, platforms, and developers.

But all these developments only scratch the surface of what an individual creator can achieve in the creator economy of the future, considering the promise of creator-driven decentralized autonomous organizations, or "creator DAOs" (which will be discussed in Chapter 10 at the end of this book). Such new business entities are expected to be wholly owned by creators in partnership with their backers, consumers, and cocollaborators – with their financial interests aligned solely on the business of creation. The ongoing development of Roblox illustrates that the future Metaverse has tremendous potential to support new business models and content types at scale.

P2E Blockchain Gaming – GameFi with NFT

One major issue with gaming, even at user-centered UGC platforms like Roblox or Fortnite, is that gamers cannot take their money out of a game when they stop playing. Games rarely allow asset transactions between players, especially on third-party websites. Furthermore, even in games where users can trade, they cannot legally own assets because the studio has no incentive to give out their profits to users.

So, if gamers are willing to spend that much in virtual worlds without any potential return on investment, powering the gaming sector to become the largest media globally, how much bigger would the gaming community be when their in-game purchase becomes an asset that can be traded and used for real-life transactions? This is where the potential of blockchain gaming is and why games could drive wider blockchain adoption.

The best example in the current market is play-to-earn (P2E) games, where DeFi and NFTs (covered in previous chapters) are integrated into gaming. The combination of DeFi, NFTs, and games have enabled players to be financially rewarded for playing the games, which is also referred to as *GameFi* (**see Figure 6.3**). Game players can use various methods to monetize NFTs obtained in the game to make a profit. Vice versa, NFT gains value from the game.

Behind a GameFi is a carefully designed token economics system with incentives balancing NFTs, which usually represent some sort of in-game assets and a utility token that represents a sort of internal currency. Generally game assets in the form of NFTs are characters used in the game, virtual terrain, and in game items like weapons. Players use these NFTs to earn the utility token, which has a value on the market and can be exchanged via DeFi infrastructure into other cryptocurrencies or real cash.

The GameFi protocols represent the prime watershed moment for NFTs that followed the Metaverse narrative. Before

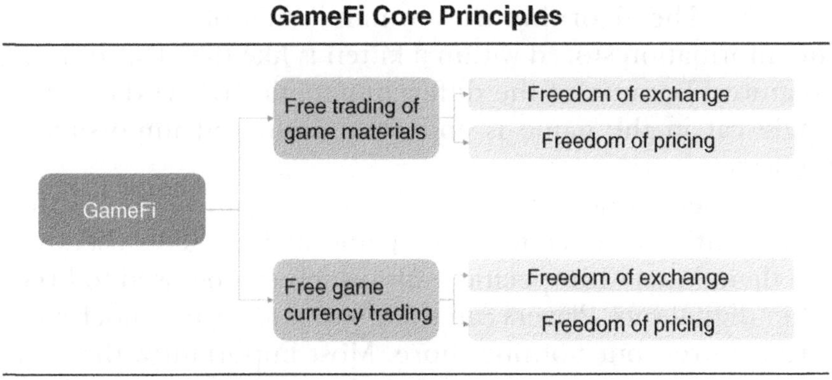

GameFi Core Principles

Figure 6.3   GameFi Empowers Free Trading of In-Game Cryptos
Source: CoinTelegraph, Medium, 2021

that, the most popular category of NFTs had been collectibles, according to Nonfungible.com's research in early 2022. But the collectibles, like CryptoPunks and Bored Ape Yacht Club, are the most basic type of NFTs. (See related discussion in Chapter 5.) Having the popular collectibles means the holder is a serious collector, and using them as the users' avatar could help the users to enter some more advanced communities (as a club membership card), and they may get free airdrops from other NFTs (such as Meebits).

However, except for speculation and influence, such "collectible NFTs" have no other use. By integrating with gaming, NFTs have evolved into more complex and sophisticated versions of interactive NFTs. We can divide these NFT developments into three different levels.

Level 1: Basic Interaction NFT (Example: CryptoKitties)

Compared to the simple collectible NFTs, CryptoKitties provides an extra gamification component: reproduction. It is a game where you collect, breed, and even sell virtual cats for real money. Blockchain is the technology underpinning the game, which runs on an algorithm by the name of the *genetic algorithm.* The algorithm tries to mimic that of real genetics – the information stored within a kitten is like the DNA of living creatures. Because of the different genetic (DNA) data, every single cat in the game is entirely unique and impossible to replicate.

Still, being one of the first blockchain games ever created, CryptoKitties is a "collectable" game at best. The users collect digital cats with specific traits, which can be used to breed other digital cats. Players can trade cats and try to unlock some rare features, but nothing more. Most importantly, there are almost no real games at CryptoKitties, and players could find no meaningful uses for the digital assets they collect.

Level 2: Explore (Examples: Decentraland, Sandbox)

Decentraland and Sandbox represent the development of level 2 games: virtual platforms that allow users to create, experience, and monetize content and applications. The two virtual worlds are similar in that they are both decentralized virtual world platforms where users can buy and sell virtual plots of land, and freely create and design virtual worlds or experiences on the land. Both are based on the Ethereum blockchain. (Cryptovoxels, inspired by Minecraft, is another virtual world/metaverse powered by the Ethereum blockchain, allowing players to buy land and build stores and art galleries.)

Decentraland is an NFT game that presents players with a bounded 3D virtual world, where the LANDs are situated adjacent to one another, and they are the NFTs of this game. Multiple LANDs can be grouped and formed into communities called districts. Players with shared interests would create their own districts for multilateral benefits between each other.

The Ethereum-based Sandbox is effectively a playable chunk of the Metaverse, providing a shared online world in which users can purchase NFT land plots, create their own interactive games on them, and even monetize those experiences as they share with other users. The Sandbox is made of three products: (1) VoxEdit, which allows users to create and animate 3D objects in the Sandbox metaverse, such as people, animals, and tools (referred to as ASSETS, which can be fungible or nonfungible), (2) the Sandbox Marketplace, which allows users to publish and sell their ASSETS; and (3) the Sandbox Game Maker, which allows users to create 3D games for free. To encourage a broader base of users to participate in the content creation process, no coding experience is required as users can use Sandbox's own platform's tools to create ASSETS and games.

Virtual Land Booming

For average people, it may seem odd why people now are purchasing real estate with real dollars (or cryptocurrencies) in a virtual gaming setting, such as Decentraland and the Sandbox.

Decentraland is a prominent example of how virtual real estate can boom in the metaverse era. Besides landmark sales valued over millions of dollars (e.g., real estate company Metaverse Group's $2.43 million purchase of Decentraland land), governments are also buying into the action. In November 2021, Coindesk reported that Barbados had partnered with Decentraland to build its virtual embassy on the platform in January 2022, making Barbados the very first country to build a virtual embassy. Decentraland has thus succeeded in not only gaining popularity among users (300K monthly active users, Dec 2021), but has now also gained traction on the institutional business and government side.

The Sandbox's virtual real estate sale broke the record set by Decentraland. According to the *Wall Street Journal*, Republic Realm purchased $4.3 million's worth of land in the Sandbox from Atari SA, which was the largest metaverse property sale to date as of November 2021. Yat Siu, co-founder of Animoca Brands (of which the Sandbox is a subsidiary), voiced that: "We don't think the Sandbox will be the only place, but it's one of the first places that has become kind of like the digital Manhattan or the digital Beverly Hills." Indeed, if one was given an opportunity to purchase real estate in what would be Manhattan's prime midtown, or Beverly Hills' shopping arcades, it would explain the record-shattering $4.3 million price tag.

Virtual land sales are booming globally, with China being no exception. One prominent example is artist Huang Heshan's metaverse art project, TooRichCity. TooRichCity, according to Sixth Tone, is a virtual city project composed of huge, tottering towers made of 3D concrete and rustic, campy shop signs

Huang collected in China's lower-tier cities and villages. By Jan 2022 (time of publication of the Sixth Tone article), Huang had sold 310 of his NFT houses within two days for 400,000 yuan ($63,000). The buyer demographic tends to be younger, and each received a property certificate and an invitation into a WeChat group for NFT property owners.

These high sales for digital renderings of real estate are particularly intriguing. For one, unlike physical property owners, TooRichCity homeowners will not be able to walk into their properties or take shelter. For another, these sales of virtual real estate come at a time when physical real estate in China is in shambles in the wake of increasingly stringent regulations. These factors, however, do not seem to dampen the enthusiasm of Chinese investors.

Three major firms have driven the development of the hype around virtual real estate: Meta (Facebook), Decentraland, and the Sandbox have all given rise to user and investor enthusiasm around metaverse properties. Traditional real estate heavyweights have also entered the ring, with New World Development and Sun Hung Kai Properties investing in the Sandbox.

Decentraland, comparatively, is more centered on the creation and ownership of assets such as buildings, gardens, transport systems, and various other types of architecture. According to PrestigeOnline, there are 90,601 individual plots of virtual land on the Decentraland system. Once a plot of virtual land is purchased using MANA, it can be used to create a small garden or an entire city by the user. Decentraland's real estate operates along rules that are similar to that of physical real estate, in that land groups that are similar to each other can be grouped as a district, fostering communities of likeness. There is also a voting system (Agora) that operates similar to housing board voting systems, where users can vote and influence developments in their communities. (See **Box: Virtual Land Booming**.)

Level 3 P2E Gameplay (Example: Axie Infinity)

Axie Infinity leads the P2E (play-to-earn) gaming trend. Axie Infinity is a digital pet universe where players battle, raise, and trade fantasy creatures called *Axies* in the form of NFTs. Axie

Axie Infinity's Economic System

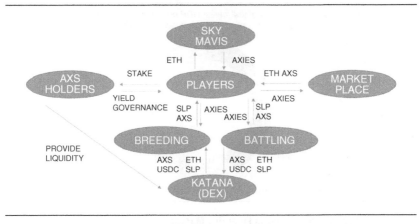

Figure 6.4   Axie Infinity's P2E Ecosystem
Source: TeehFlow

users start the game by investing in Axie NFTs and AXS native tokens, then they can collect, breed, fight, and trade the Axies creatures. Players can also earn in-game currency (Smooth Love Potion, or SLP) by selling earned Axie tokens for other cryptocurrencies or fiat currencies (see **Figure 6.4**). Despite still being in its early stages, Axie Infinity has made a resounding name for itself by being ranked #1 Ethereum game from daily, weekly, and monthly community voting.

Axie Infinity is now one of the world's fastest-growing video games, and the game has surpassed more than $1 billion in sales with over one million daily users. Interestingly, this P2E game has been especially popular in emerging markets. For example, during the economic difficult period caused by Covid-19, many users in the Philippines earn more than their usual monthly salary just by playing Axie Infinity, and in 2021 the Philippines contributed the most user traffic for the game among all countries (see **Figure 6.5**). P2E gaming may become a catalyst needed to take the blockchain game sector to the next frontier globally.

Axie Infinity Web Traffic by Country

*Countries among global lowest GDP per capital make up
more than 50% of all Axie Infinity website traffic*

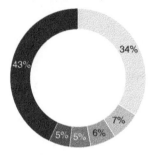

Philippines ▨ Venezuela ■ Argentina ■ Brazil ■ United States ■ Others

Figure 6.5   Axie Infinity Popular in Emerging Markets

Source: SimilarWeb.com, Ciypto.com Research; Data as of 23 Nov 2021

Blockchain Gaming: Gaming First? Crypto First?

Since mid-2021, we have witnessed the first outbreak of crypto games. Even well-known streamers and gamers have started to switch to NFT gaming, and big names in the game industry, such as Activision, Blizzard, EA Sports, Ubisoft, and Square Enix, have begun dipping their toes into blockchain and NFT -based games. And yet, the blockchain gaming market is currently relatively small, and most development has been led by crypto native companies, who are not gaming native.

In addition, the traditional gaming industry has not exactly been receiving this innovation well. As the number of NFT announcements from game studios accumulated players became increasingly annoyed, and clashes over crypto have increasingly erupted between users and major game studios like Ubisoft, Square Enix, and Zynga. In many of the encounters, the gamers have prevailed – at least for now. In short, the video game world is divided over NFTs.

For example, Ubisoft, which makes titles like Assassin's Creed, was the first large game publisher to wade into crypto. In December 2021, Ubisoft debuted a platform called Quartz, which lets players own in-game cosmetic items (like helmets and guns) in the form of NFTs. The NFTs were available for free in the shooter game Ghost Recon Breakpoint for players who had reached a certain level in the game. Gamers, the company said, could keep the items or sell them on third-party markets. The move was met with widespread anger from gamers, who slammed Quartz as a cash grab. A YouTube video about the move was disliked by more than 90 percent of viewers.

Furthermore, some major game companies have come out against crypto. Valve, which owns the online game store Steam, updated its rules in late 2021 to prohibit blockchain games that allow cryptocurrencies or NFTs to be exchanged. Epic Games said its company would steer clear of NFTs in its own games because the industry is riddled with "an intractable mix of scams." (Epic will still allow developers to sell blockchain games in its online store.)

To summarize, it's fair to say that right now blockchain games are still in an embryonic stage. On one hand, for **creators and developers**, blockchain gaming is poised to demonstrate the following key benefits:

- **Better game economies.** P2E creates an opportunity to monetize more players vs. the free-to-play model where, on average, less than 2 percent of players purchase in-game items (appier data in 2020). P2E may engage all the players in a game into economic transactions.
- **Better unit economics.** Sharing a portion of economics with players and creators in the game economies can result in lower customer acquisition costs and greater retention, contributing to higher LTVs (lifetime value) per user than traditional free-to-play games.
- **Better economic alignment.** With gaming platforms like Roblox (see related discussion earlier in this chapter), creators keep approximately 30 percent of revenues. With blockchain games, users generally retain much more of the value they help create. Furthermore, by leveraging

things such as on-chain royalties, developers can unlock a new revenue stream by collecting fees on secondary market activity.

On the other hand, **gamers** worry that the addition of cryptos like NFTs will allow games to become a platform just for commerce. What they love about games is the worlds, stories, and experiences inside them; what they don't want is for things to feel transactional in the games.

Therefore, in terms of the quality of blockchain games, an important test for developers would be to examine a blockchain P2E game as if it were a traditional free-to-play game with no token-earning potential. If it could be successfully monetized through microtransactions (i.e., gamers are willing to pay for in-game items with no expectation of financial return), then the gameplay is sufficiently engaging.

However, if it would not make much through microtransactions – players would not pay for assets without earning potential – then the only reason they would pay with P2E is as an investment. If that's the case, it is not standalone gaming. This makes the game zero-sum and leads to a pump-and-dump scenario. It is more of an investment vehicle with a game wrapper than an actual game. In other words, it is a "crypto first" gaming, and such gameplay cannot stand on its own.

To conclude, blockchain gaming should be quality gaming to begin with, with the additional crypto play as a valuable enhancement. If the future blockchain gaming can be "gaming first" instead of "crypto first," the GameFi revolution will inevitably occur among the gaming communities and bring the next billion users into the crypto field.

Gaming, the Foundation of Metaverse

Before concluding this chapter, one important question: Why is the gaming revolution critical for metaverse development?

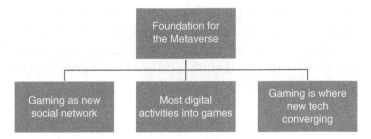

Figure 6.6   Gaming Is the New Technology Paradigm

Metaverse is partly a dream for the future of the internet and partly a neat way to encapsulate some current trends in online infrastructure, including the growth of real-time 3D worlds. Gaming is a test case for the Metaverse, considering that most digital activities are migrating to gaming, and it is also the new social network.

Since 2021, gaming – now an industry valued to exceed $200 billion – stands as the largest media category by revenue. Having recently taken the leading position from long-term media behemoth linear television, gaming today is larger than the global music, film, and on-demand entertainment sectors combined. Furthermore, among these media categories, gaming is also the fastest growing, according to various industry research firms. Therefore, gaming is the *best* test case.

Further considering that gaming is where the new technologies are converging, we may even say that gaming is the new technology paradigm (see **Figure 6.6**). In short, games are the starting point and the most viable path toward the Metaverse, because:

- Many games are already at scale (e.g., massive, engaged user bases).
- UGC culture is from the creator side (e.g., user-generated activities, games, virtual goods, environments/worlds).

- From the consumption side, games already provide experiences for consumers to participate in metaverse-like activities (e.g., using NFT for digital assets, online-merge-offline, real-time connection between physical and digital experiences).
- Technologies to build a metaverse are already being developed and tested in gaming (e.g., AR/VR integration, multi-gamer concurrency, content moderation).

So far in Part II, **Chapters 3 and 4** focused on blockchain tech as the infrastructure for users to "transact" in the Metaverse; **Chapters 5 and 6** mostly focus on the creator economy, illustrating how blockchain technologies enable the digital content and asset creation. In the following two chapters, we will focus on the blockchain foundation for user data privacy and crypto assets security in the Metaverse.

CHAPTER

7

Metaverse Privacy: Blockchain vs. Big Tech

- Privacy in a Parallel Digital Universe
- Future Data Privacy Model in Metaverse
- WEF Data Governance Model
- Zero-Knowledge Proof and Secure Multiparty Computation
- Homomorphic Encryption and Federated Learning
- NFT "Cookies": When Web3 Tech Meets Web2.0 Legacy
- Surveillance Economy and Dystopian Society

Privacy in a Parallel Digital Universe

Facebook's "Meta push" in 2021 was met with at least as much suspicion and hesitancy, especially relating to data privacy concerns, as the enthusiasm it has received. The whistleblower Frances Haugen, in an interview with the Associated Press, said the metaverse world could give Facebook another monopoly online, as well as being addictive, and steal even more personal information from users. Haugen said Facebook's recent trumpeting of the Metaverse is a screen behind which Facebook (now Meta) can hide while its regulatory issues play out: "If you don't like the conversation, you try to change the conversation," she said.

Meanwhile, Apple's tracking IoT device AirTag has also revealed some privacy issues. Despite AirTag's built-in privacy features, some observers pointed out that Apple has not gone far enough to protect people from unwanted tracking. One *Washington Post* reporter allowed a colleague to track him with an AirTag for a week – before putting a stop to it. A December 2021 report from *The New York Times* contained reports from at least seven women who believed they were tracked by AirTags.

Also in December 2021, police in Canada issued a warning that thieves were using the Apple tracking accessory in the theft of high-end vehicles. Specifically, they had five reports of possible AirTag involvement, out of more than 2,000 reports in total. However, that one specific quirk of AirTags isn't the only difference between the Apple tracking accessory and other products. Furthermore, a larger platform for tracking is the LTE telecom network itself, which is leveraged by hundreds of standalone products, priced similarly to AirTags.

Privacy and antitrust are intertwined. In January 2022, Microsoft announced that the company was acquiring Activision Blizzard, the gaming giant responsible for such mega hits as Overwatch, Diablo, Call of Duty, World of Warcraft, and Candy Crush. The nearly $70 billion acquisition of the recently troubled gaming studio speaks to the growing trend of large tech companies buying and merging with smaller ones to consolidate power in the tech industry.

While many high-profile members of the Biden administration have called for greater regulation over tech to curb antitrust and anticompetitive practices, the concentration in the tech industry also poses a privacy problem. Microsoft already owns the popular Xbox console platform and profitable gaming franchises like Halo, Forza, Age of Empires, and Minecraft. The Activision Blizzard deal will likely be even more impactful, since this is the second major game studio Microsoft has purchased in less than a year.

In 2021, Microsoft acquired ZeniMax, the parent company of popular games studio Bethesda Softworks, in a $7 billion deal. It's clear that Microsoft is moving to take over a large part of the gaming market to aim at a universal metaverse platform for play, work, and social. Microsoft's announcement noted that the acquisition would "provide building blocks for the metaverse." The Activision Blizzard deal will also mean a lot more user data for Microsoft (and it already has access to quite a lot). The privacy risks related to data collection are compounded when a company can collect vast amounts of data from different sources, all on the same individual. It becomes easier for that data to fall into the wrong hands or lead to misuse of private data.

For example, these large platforms are able to profile their uses in striking details. As the "mosaic theory" suggests, disparate items of information, though individually of limited or no utility to the owner, can take on added significance when combined with other items of information. In the cyberspace, there is a lot of different information that a user would never think would be able to identify a person. But when a computer combines the different pieces together, the computer can see connections in ways that humans cannot. When a digital platform combines different sets of data, either from different service lines of the same platform or from third-party data vendors in different sectors, the power of user data integration grows exponentially.

The larger and more powerful a tech company becomes, the more it is able to collect and use data from private individuals. Likewise, the more data a company has, the harder it is for other companies to compete. On one hand, the platforms have used big data analysis to provide users with more personalized services and faster services. If one company has access to billions of data points on user behavior and preferences, it is likely more able to meet consumer demand than a smaller company that lacks those resources.

On the other hand, the data power of many platforms has also aroused the public concerns that big data could be abused. In the case of "big data killing," the more personal data the platforms have, the more users have to pay. (See **Box: Big Data Killing**.) Again, data privacy and antitrust are intertwined. The concentration of data power hurts consumers and tech startups.

Big Data Killing

Shashu literally means "killing someone that a person is acquainted with," and it is a term coming out of China's market economy, which refers to the situation where a person takes advantage of another who innocently believes the person is a friend and acts in his or her interest.

In the data economy it evolves into "big data killing," as the internet platforms capitalize on their regular users when their spending habits are well known by the platforms. The "killing" refers to the situation where for the same goods and services, the price shown to old customers is more expensive than that to new users. In economic term, "big data killing" is a form of price discrimination.

Big data killing illustrates the power and value of data. Because the internet platform has no knowledge about the new user, it would offer a relatively low product (or service) price so that the new user can enjoy the "sweetness" of the first experience (at the same time, the platform gathers his or her personal data through the user's platform registration and related transactions). Meanwhile the internet platform offers a relatively higher price to existing users, especially those who are analyzed to have high spending power and low sensitivity to pricing.

For example, online travel agency (OTA) sites are where big data killing is prevalent. Many users have discovered that when they try to book air tickets or hotel rooms, the price would be higher for a frequent user of the website than a newcomer. Online car-hailing platforms are also found to offer different prices in the same region to different users. Similar phenomenon is also reported to occur in online shopping, online ticket purchases, video websites, and many more fields.

Going forward, consumer data is going to be at the very heart of the Metaverse, and Big Tech companies' extensive data gathering in metaverses will become an even bigger data privacy issue. As a 3D, persistent, immersive, and interactive internet, the Metaverse presents an opportunity to translate everyday activities – working, attending a concert, traveling, shopping, socializing – into a parallel digital universe, which is sure to rely on new kinds of potentially sensitive information like biometric data.

Metaverse applications could let tech companies track your facial expressions, blood pressure, your breathing rates, and even more. For e-commerce businesses, that's great news, because brands can, based on extensive user data, create ads that will be immersive, personalized, and value-based experiences. But for personal privacy, there is little for the user to hide. For example, A VR/AR headset alone could serve as both camera and microphone inside of homes; more advanced VR systems could pair this with heart and respiration rates, physical movements, and 3D dimensions, and use the unique combinations of all of this information for individual identification and tracking.

Therefore, with the dawn of the Metaverse upon us, debates about data-driven advertising and consumer data protection are already heating up. The Metaverse may feed Big Tech companies' appetite for more data, and their centralized stored and controlled data will pose more challenges to privacy, unless the data can be decentralized, and the control and ownership is returned to the individual users. In this chapter, we will discuss a few data governance models for metaverse applications, as well as new technologies that can be integrated to provide better privacy protection.

Future Data Privacy Model in Metaverse

There are four potential data privacy governance models in the Metaverse, ranging from tightly central-controlled ownership to individual ownership (see **Figure 7.1**).

Centralized Governance Model

In a centralized governance model, the owner and operator of that environment sets the policies by which the service is governed. Users' data doesn't leave that operator's metaverse, and it is not interoperable with other metaverses (for example, those developed by different major tech companies) unless connections are enabled on the metaverse's owner and operator's terms.

This model is most akin to the current model adopted by Big Tech, such as Meta, Amazon, and Twitter. The individual user privacy in this model is least protected because big companies have centralized control of the data, and they also have sophisticated algorithms to make use of the individual users' data. Furthermore, the hackers can also take advantage of the centralized nature of data to steal data. This certainly is not ideal for metaverse applications.

Individualized Model

An individualized based governance model in the Metaverse is the idea that once the technology is more commoditized, anyone could build their own metaverse and connect it loosely to those of other creators. This model is most akin to the "open web" notion of digital publishing where access is not restricted, standards for publication are minimal if they exist at all, and data can move freely between metaverses.

In such an "open metaverse," the individual users are also creators of their own metaverse, and the privacy can be protected based on the necessary consent model. Many experiments are happening (e.g., in the gaming industry), as illustrated in multiple sections of this book. This is an ideal privacy governance model.

DAO-Based Governance Model

Individual governance model gives maximum and ideal protection for user's privacy, but it is hard to implement currently. (It can be viewed as the ultimate direction of the metaverse

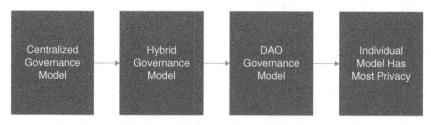

Figure 7.1   Different Degree of Privacy Protection of Four Governance Models

from both technology and regulatory perspectives.) However, as in any "network effects" product, most users will still consolidate around relatively few different metaverses, with the exception of some "long-tail" metaverses for those with niche interests.

Therefore, if we have no mandated standards, the data privacy and security infrastructure practices in the popular metaverses may well end up being worse than those that would have been supported by a centralized approach. One potential solution is DAO (or decentralized autonomous organization) based governance model, where DAO participants leverage governance tokens to vote for privacy and other DAO related issues.

One such example is Panther Protocol, which was launched in early 2022. Panther provides DeFi users with fully collateralized privacy-enhancing digital assets, leveraging crypto-economic incentives, and zkSNARKs (one of the privacy persevering computation algorithms discussed later in this chapter) technology. Users can mint zero-knowledge zAssets by depositing digital assets from any blockchain into Panther vaults. (Zero-knowledge proof is another privacy persevering computation algorithm that will be introduced later in this chapter.)

Panther deployed LaunchDAO, a system allowing every user that had completed KYC identity verification for its public and private token sales to issue a zero-knowledge proof anonymously verifying their participation. Using this proof, individually verified users could privately vote on whether to launch the protocol on the Ethereum and Polygon blockchains.

The DAO approach to privacy is still in the early innovation stage, and where it will be successful will depend on multiple factors, among them the most important factor is the regulatory aspect. A hybrid governance model that balances the government regulatory consideration and the needs of individuals (and DAO) may be the best approach we can have.

Hybrid Governance Model

The "hybrid" option allows for creators, Big Tech, and DAO participants to set their own governance policies; however, key data privacy, community, and security standards are governed by the sovereign states (needs to meet the local privacy regulations). The chief standard among these would be to require firms hosting metaverses to enable data portability and interoperability between other metaverses so as to ensure consumers are not "locked in" to one platform. This forces metaverse providers to compete on quality and services, rather than relying on high switching costs, to keep users in their networks.

The hybrid model represents an opportunity for the state to provide more effective data privacy governance (Web3) than we experienced at the onset of online digital networks (Web2.0). The hybrid model still benefits from the privacy and security infrastructure expertise of what we assume will be the major metaverse firms like Facebook and Amazon – but it also takes away the potential for total monopoly power over user interactions in the Metaverse. The portability and interoperability requirements – from the sovereign regulators – will be crucial to this. However, portability and interoperability are not enough to fully protect privacy, and we must also enforce policy requirements that dictate what metaverse providers can do with data collected in their worlds.

WEF Data Governance Model

World Economic Forum (WEF) in its white paper titled "Data for Common Purpose: Leveraging Consent to Build Trust" (published November 2021) proposed a consent-based privacy and trust model for data exchange and governance. This model can be used for the metaverse platform as a reference model. The white paper defined 16 attributes (see **Figure 7.2**) for building trust through consent mechanisms in data exchanges for the common good. We can view the Metaverse as the ultimate platform for data exchanges, and the WEF model shall fit well, too.

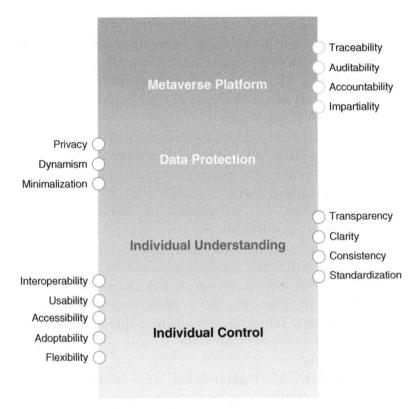

Figure 7.2   Metaverse Privacy and Governance Reference Model
Source: World Economic Forum (WEF)

The 16 attributes are organized into four different groups: Metaverse Platform Accountability, Data Protection, Individual Understanding, and Individual Control.

Metaverse Platform Accountability

Measuring, tracking, and reporting how data moves through and outside of the data exchange or any centralized data platforms. The following four attributes are defined:

- **Traceability.** The ability of metaverse platform stakeholders to follow their data from consent, through collection, use, and sharing, to termination – including the first and nth generation of consumption levels. The current technical limitations need to be recognized and governance designed to accommodate future approaches.

- **Auditability.** Metaverse platform operators and participants should maintain, for a period equal to the respective jurisdictional retention period for recourse of data misuse or other requirements a record of: (1) each instance of data collection, use, or sharing; and (2) consent choices for data use explicitly provided by the individual to support audits.

- **Accountability.** Any participant or metaverse platform operator is accountable for protecting individuals' privacy, providing methods for recourse, and using their data only within the consent parameters they or a third party who is legally acting on behalf of the individual (e.g., parent, guardian, data cooperative, trusted agent) have chosen. When at all possible, this extends to any additional processing of data outside of the exchange participants.

- **Impartiality.** All stakeholders who participate in a metaverse platform, regardless of entity size, stature, or other potential biases, should be held equally accountable for the actions they take within a data exchange.

Data Protection

Protecting data and reducing risk is associated with the data stored within a data exchange (i.e., data at rest), data being processed (data in use), and data flowing within, to, and from (i.e., data in motion) a metaverse platform. The following three attributes are defined for data protection:

- **Privacy.** Individuals should be provided with consent-management tools and experiences to control the collection, use, and sharing of personally identifiable (PI) data and data of personal origin, in coordination with the rights afforded to individuals through data privacy policies within their jurisdictions. Metaverse platform operators are responsible for enforcing these user controls.

- **Dynamism.** Metaverse platform operators provide participants with governance rules and enforce an individual's ability to grant, modify, and revoke consent for data collection, use, or sharing. In scenarios where consent modification and revocation are limited, such limits are communicated to users in advance of the initial permission experience.
- **Minimization.** Individuals' data collection, use, and sharing should be done in accordance with their consent choices, ensuring they are strictly relevant and adequate for a specific purpose. Metaverse platform participants enable the necessary tracing and audit capabilities for operators to ensure compliance.

Individual Understanding

This group focuses on educating individuals and positioning them to make informed choices on how data is collected, used, and shared through a metaverse platform. The following four attributes are defined for individual understanding:

- **Transparency.** Individuals can access clear information in due time and with ongoing visibility on how an entity collects, uses, and shares their data, the potential risks and benefits of the service, the availability of source code, the rules and standards upon which it is based, their rights and obligations, and the governance structure of the metaverse platform.
- **Clarity.** Individuals are enabled to have a clear understanding of how and why data is collected, used, and shared. Consent choices are presented in a coherent and intelligible way. No dark-pattern behavior is permitted.

- **Consistency.** Consent mechanisms are presented to individuals in the same format across all user interactions in the data exchange (e.g., terminology, icons, structure).
- **Standardization.** Consent choices are presented to individuals, across use cases and participants, in a common, digestible format. Participants enable the necessary technology to read, interpret, and implement consent choices received in the standard format. Operators enforce these standards.

Individual Control

This group addresses supporting the tools and interfaces for individuals to exercise their informed choices on how their data is collected, used, and shared through a metaverse platform. The following five attributes are defined for the individual control:

- **Interoperability.** Transfer capabilities for individuals' data, along with the associated rights and permissions, from one operator or participant to another operator or participant, are in a common machine-readable format that is interpretable and understood across participants in a metaverse platform.
- **Usability.** Individuals are able to control how their data is used, collected, or shared. They are enabled, through the data exchange consent mechanisms, to act to control the collection, use, and sharing of personally identifiable data as well as de-identified data of personal origin.
- **Accessibility.** Individuals are able to control how data is collected, used, and shared through a simple consent-mechanism interface that is available at relevant interaction points. Access should be available for all individuals, including minorities, those with disabilities, and individuals of low socioeconomic status.

- **Adoptability.** The governance approach for individuals' interactions with consent mechanisms should be implementable and manageable by both operators and participants in a metaverse platform.
- **Flexibility.** Consent interfaces allow for easy modification to respond to the altered circumstances of operators, participants, or individuals, as well as governance rules, regulatory policy, or technology developments.

Zero-Knowledge Proof and Secure Multiparty Computation

To protect users' privacy in terms of data, identity, digital assets, and transactions, there are a few privacy preserving technologies currently under active research. The following two sections will introduce four of the privacy technologies that can be heavily leveraged by metaverse platforms (see **Figure 7.3**).

Zero-Knowledge Proof

What Is Zero-Knowledge Proof?
Is it possible to show that something is true without revealing the data that proves it? This is what zero-knowledge proof (ZKP) technology proposes – a method for one party to cryptographically prove to another that they possess knowledge about a piece of information without revealing the actual underlying

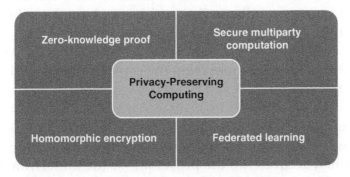

Figure 7.3   Leading Privacy-Preserving Computing Methodologies

information. In the context of blockchain networks, the only information revealed on-chain by a ZKP is that some piece of hidden information is valid and known by the prover.

ZKPs were first described in a 1985 MIT paper from Shafi Goldwasser and Silvio Micali called "The Knowledge Complexity of Interactive Proof-Systems." In this paper, the authors demonstrate that it is possible for a prover to convince a verifier that a specific statement about a data point is true without disclosing any additional information about the data.

How Does Zero-Knowledge Proof Work?

ZKPs can either be interactive—where a prover convinces a specific verifier but needs to repeat this process for each individual verifier—or non-interactive—where a prover generates a proof that can be verified by anyone using the same proof. Additionally, there are now various implementations of ZKPs including zk-SNARKS, zk-STARKS, PLONK, and Bulletproofs, with each having their own tradeoffs of proof size, prover time, verification time, and more.

The three fundamental characteristics that define a ZKP include:

- **Completeness.** If a statement is true, then an honest verifier can be convinced by an honest prover that they possess knowledge about the correct input.
- **Soundness.** If a statement is false, then no dishonest prover can unilaterally convince an honest verifier that they possess knowledge about the correct input.
- **Zero-knowledge.** If the state is true, then the verifier learns nothing more from the prover other than that the statement is true.

At a high level, the creation of a ZKP involves a verifier asking the prover to perform a series of actions that can only be performed accurately if the prover knows the underlying information. If the prover is only guessing as to the result of these

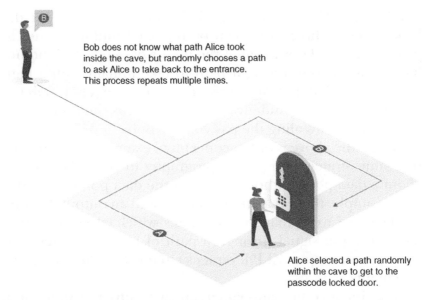

Bob does not know what path Alice took inside the cave, but randomly chooses a path to ask Alice to take back to the entrance. This process repeats multiple times.

Alice selected a path randomly within the cave to get to the passcode locked door.

Figure 7.4   Zero-Knowledge Proof Algorithm Illustration

actions, then they will eventually be proven wrong by the verifier's test with a high degree of probability.

How ZKP Works to Prove Knowledge about Data Without Revealing the Data to Another Party

A conceptual example to intuitively understand proving data in zero-knowledge is to imagine a cave with a single entrance but two pathways (path A and B, as shown in **Figure 7.4** – you may recall Bob and Alice appeared in Chapter 2 earlier for a bitcoin transaction, see **Figure 2.2**) that connect at a common door locked by a passphrase. Alice wants to prove to Bob she knows the passcode to the door but without revealing the code to Bob. To do this, Bob stands outside of the cave and Alice walks inside the cave, taking one of the two paths (without Bob knowing which path was taken). Bob then asks Alice to take one of the two paths back to the entrance of the cave (chosen at random).

If Alice originally chose to take *path A* to the door, but then Bob asks her to take *path B* back, the only way to complete the puzzle is for Alice to have knowledge of the passcode for the locked door. This process can be repeated multiple times to prove Alice has knowledge of the door's passcode and did not happen to choose the right path to take initially with a high degree of probability. After this process is completed, Bob has a high degree of confidence that Alice knows the door's passcode without revealing the passcode to Bob.

While the above is only a conceptual example, ZKPs deploy this same strategy but use cryptography to prove knowledge about a data point without revealing the data point. With this cave example, there is an input, a path, and an output. In computing there are similar circuit systems, which take some input, pass the input signal through a path of electrical gates, and generate an output. ZKP leverages circuits like these to prove statements.

Imagine a computational circuit that outputs a value on a curve, for a given input. If a user can consistently provide the correct answer to a point on the curve, one can be assured that the user possesses some knowledge about the curve since it becomes increasingly improbable to guess the correct answer with each successive challenge round. One can think of the circuit like the path that Alice walks in the cave; if she is able to traverse the circuit with her input, she proves she holds some knowledge, the "passcode" to the circuit, with a high degree of probability.

Benefits of ZKPs

Being able to prove knowledge about a data point without revealing any additional information besides knowledge of data provides several key benefits, especially within the context of blockchain networks. The primary benefit of ZKP is the ability to leverage privacy-preserving datasets within transparent systems such as public blockchain networks like Ethereum.

While blockchains are designed to be highly transparent, where anyone running their own blockchain node can see and download all data stored on the ledger, the addition of ZKP technology allows users and businesses alike to leverage their private datasets in the execution of smart contracts without revealing the underlying data. Ensuring privacy within blockchain networks is crucial for preserving privacy in various metaverse applications.

Secure Multiparty Computation

What Is Secure Multiparty Computation?
Secure multiparty computation (also called multiparty computation, SMPC, or MPC) is a cryptographic technique that enables different parties to carry out a computation using their private data without revealing their private data to each other. Secure multiparty computation is a method that can help businesses and individuals ensure the privacy of their sensitive data without undermining their ability to gain insights from it.

How Does Secure Multiparty Computation Work?
A popular example to illustrate the basic idea behind SMPC is as follows: Suppose a group of employees wants to learn their average salary in order to find out whether they are underpaid. However, they don't want to disclose their individual salary information. An SMPC method can solve this problem with the following steps:

1. Each employee is numbered from first to last.
2. The first employee chooses an arbitrarily large number and adds their salary to the number and tells the second employee the result.
3. The second employee adds his or her number to the value and tells the result to the third employee, and so on until the last employee.
4. After adding their salary to the result, the last employee tells the result to the first employee.
5. The first employee subtracts the large number they started with and divides the result by the number of employees in the group to obtain the average salary.

In this example, the large number chosen by the first employee hides his/her salary from the others. At the same time, the final result that the first employee receives from the last employee provides no information to the first employee about the others' salaries. As a result, the group, consisting of multiple parties, could securely compute the average salary without disclosing their salaries.

This is, of course, a simple example to illustrate how SMPC works. In real-world use cases, SMPC enables complex computations such as machine learning models using privately held data without the need of sharing it.

What Are the Properties of Secure Multiparty Computation?
SMPC aims to ensure two basic properties against adversarial attacks:

1. **Input privacy.** No party can infer information about private inputs from the output.
2. **Correctness.** An adversarial party must not be able to prevent other parties from receiving their correct outputs.

An adversary in this context refers to the parties that attack the computation process. The attack may be for the purpose of learning private information of other parties or for causing the output of the computation to be incorrect.

What Are the Benefits of Secure Multiparty Computation?

- **Promotes privacy and data utility.** SMPC can eliminate the tradeoff between data privacy and data utility since private or encrypted data does not need to be shared with third parties or model owners to be utilized. As a result, it also eliminates the risks of data breaches and misuses stemming from data collection.
- **Reveals only the final result.** SMPC only reveals the final result and does not reveal intermediate information during the computation.

What Are the Challenges to Secure Multiparty Computation?

- **Communication overhead.** As illustrated in the example above, the SMPC method requires communication between parties, which can lead to high communication costs.
- **Vulnerable to attacks from colluding parties.** For instance, the second employee and the fourth employee can collude to learn the third employee's salary by subtracting the value sent by second to third from the value sent by the third to fourth.

Homomorphic Encryption and Federated Learning

Homomorphic encryption (HE) and federated learning (FL) are two different but related technologies that aim to solve the same problem: How can computation tasks such as machine learning and blockchain transactions be performed more privately and securely? In other words, how can an IT system use data without seeing the data?

Homomorphic Encryption

What Is Homomorphic Encryption?

Homomorphic encryption (HE) is a type of encryption method that allows computations to be performed on encrypted data without first decrypting it with a secret key. The results of the computations also remain encrypted and can only be decrypted by the owner of the private key. Homomorphic encryption enables metaverse platforms to process and transact on user's encrypted data without decryption of the data.

How Does It Work?

The process starts with data in its decrypted form (i.e., plain text). The data owner wants another party (such as cloud or metaverse service provider) to perform a mathematical operation (e.g., some function, or a machine learning model) on it without revealing its content.

As show in **Figure 7.5**, the data owner (or the client of the cloud) encrypts the data and sends it to the other party (the cloud server). The cloud server receives encrypted data, performs operations on it, and sends the encrypted result to the owner. The owner of the data decrypts it with a private key and reveals the result of the intended mathematical operation on the data.

There are three main types of homomorphic encryption:

1. **Partially Homomorphic Encryption (PHE).** PHE only allows selected mathematical functions to be performed on encrypted data.
2. **Somewhat Homomorphic Encryption (SHE).** SHE allows a limited number of mathematical operations up to a certain complexity to be performed, for a limited number of times.
3. **Fully Homomorphic Encryption (FHE).** FHE allows any kind of mathematical operation to be performed for an unlimited number of times.

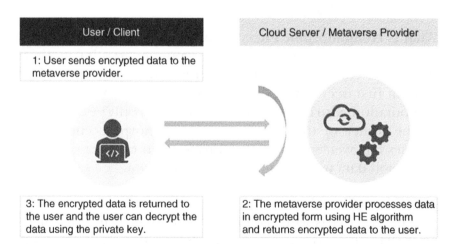

Figure 7.5   How Homomorphic Encryption Works

Why Is Homomorphic Encryption Important Now?

Sharing private data with third parties, such as cloud services or metaverse platforms, is a challenge due to data privacy regulations such as GDPR and CCPA (California Consumer Privacy Act). Failure to comply with these regulations can lead to serious fines and damage business reputations.

Traditional encryption methods provide an efficient and secure way to store private data on the cloud in an encrypted form. However, to perform computations on data encrypted by those methods, businesses either need to decrypt the data on the cloud, which can lead to security problems, or download the data, decrypt it, and perform computations, which can be costly and time-consuming. Homomorphic encryption enables businesses to share private data with third parties to get computational services securely. With HE, the cloud service or the metaverse company has access only to encrypted data and performs computations on it. These services then return the encrypted result to the owner who can decrypt it with a private key.

What Are the Benefits of Homomorphic Encryption?

- **Allows secure and efficient use in metaverse.** Homomorphic encryption can allow metaverse businesses to leverage cloud computing and storage services securely. It eliminates the tradeoff between data security and usability. Businesses don't have to rely on cloud services regarding the security of their private data while retaining the ability to perform computations on it.
- **Enables collaboration.** HE enables organizations to share sensitive business data with third parties without revealing to them the data or the results of the computation. This can accelerate collaboration and innovation without the risk of sensitive information getting compromised.
- **Ensures regulatory compliance.** HE can allow businesses operating in heavily regulated industries, such as

healthcare and finance, to get outsourcing services for research and analytical purposes without the risk of non-compliance.

What Are the Challenges to Homomorphic Encryption?
Partial and somewhat homomorphic encryption systems have existed since the late 1970s, but a fully homomorphic system that allows all mathematical operations on encrypted data was first established in 2009. In its current form, fully homomorphic encryption is impractically slow. It can be said that fully homomorphic encryption (FHE) is still an emerging technique for data security and utility. But it's a promising one, and we are likely to see faster versions of it that can be applied to a variety of use cases.

Federated Learning

In the context of machine learning, the concept of federated learning addresses the issues of data ownership and privacy by ensuring that the data never leaves the distributed node devices; at the same time, the central model is updated and shared to all nodes in the network. The copies of machine learning models are distributed to the sites/devices where data is available, and the training of the model is performed locally. The updated neural network weights are sent back to the main repository. Thus, multiple nodes contribute to building a common, robust machine learning model iteratively through randomized central model sharing, local optimization, local update sharing, and secure model updates.

The federated learning approach for training deep networks was first illustrated by AI researchers at Google in 2016. Given the rising concerns over privacy, the main repository or server is designed to be completely blind to a node's local data and training process. The data thus resides with the owner, thereby preserving data confidentiality, which is highly beneficial for industrial, medical AI, and metaverse applications when privacy is of the utmost importance. The topology for federated learning can be peer-to-peer or fully decentralized. (**Figure 7.6** provides an example of federated learning architecture for hospital use case.)

Types of Federated Learning

Federated learning schemas typically fall into one of two different classes: multiparty systems and single-party systems. Single-party federated learning systems are called *single party* because only a single entity is responsible for overseeing the

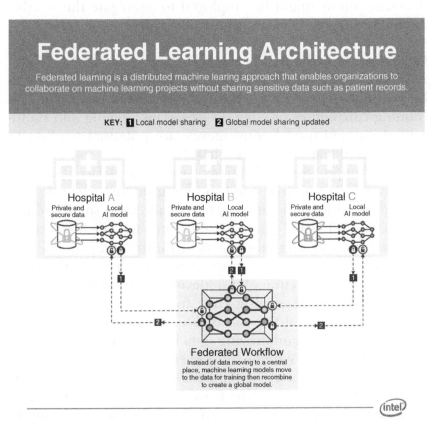

Figure 7.6   Federated Learning Architecture
Source: intel

capture and flow of data across all of the client devices in the learning network. The models that exist on the client devices are trained on data with the same structure, though the data points are typically unique to the various users and devices.

In contrast to single-party systems, multiparty systems are managed by two or more entities. These entities cooperate to train a shared model by utilizing the various devices and datasets they have access to. The parameters and data structures are typically similar across the devices belonging to the multiple entities, but they do not have to be exactly the same. Instead, pre-processing is done to standardize the inputs of the model. A neutral entity might be employed to aggregate the weights established by the devices unique to the different entities.

Challenges to Federated Learning

As federated learning is still nascent, several challenges have to be negotiated in order for it to achieve its full potential. The training capabilities of edge devices, data labeling and standardization, and model convergence are potential roadblocks for federated learning approaches.

The computational abilities of the edge devices, when it comes to local training, need to be considered when designing federated learning approaches. While most smartphones, tablets, and other IoT compatible devices are capable of training machine learning models, this typically hampers the performance of the device. Compromises will have to be made between model accuracy and device performance. Recent advancement on leveraging CPU for machine learning and training algorithms may boost device performance.

Labeling and standardizing data is another challenge that federated learning systems must overcome. Supervised learning models require training data that is clearly and consistently labeled, which can be difficult to do across the many client devices that are part of the system. For this reason, it's important to develop model data pipelines that automatically apply labels in a standardized way based on events and user actions.

Model convergence time is another challenge for federated learning, as federated learning models typically take longer to converge than locally trained models. The number of devices involved in the training adds an element of unpredictability to the model training, as connection issues, irregular updates, and even different application use times can contribute to increased convergence time and decreased reliability. For this reason, federated learning solutions are typically most useful when they provide meaningful advantages over training a model centrally, such as instances where datasets are extremely large and distributed.

Blockchain and Federated Learning
The standalone "vanilla" federated learning architecture has some limitations. The main limitations are: (1) single point of failure (the existence of central node); (2) potential lazy clients that may deteriorate the model learning; (3) no trustworthy track record of the model exchanges.

With blockchain technology, we can do the following two things:

1. **Incentive mechanism.** To build an incentive/reward mechanism to urge "lazy" clients to contribute to the local training. We can write smart contract functions named "Contribution" as well as the process followed for its implementation. The stakeholders of the metaverse platform using federated learning can all have well-defined blockchain addresses to sign transactions and interact with smart contracts. This contract, deployed on a blockchain network such as Ethereum network, guarantees that the user receives the reward when the updated local model is consumed by the central node. The central node takes into consideration the size of data provided and the number of rounds a node has participated in during a defined session. The incentive is assigned in the form of a digital token by transferring them to the addresses of the users.

2. Traceability of model exchanges. To maintain an immutable track record of the model exchanges during federated learning sessions. Integrity, security, and transparency are the basics of securing a metaverse application using federated learning. Thus, we can define a smart contract called "Federation" for aggregation strategy that will be followed during a learning session by the stakeholders of the network in terms of:

- Number of rounds
- Number of clients
- Algorithm name

On the client side, once every client performs the training of the initial model using its own local data, the generated weights will be hashed and encrypted using the SHA-256 algorithm and the hashes of the weights will be automatically saved in the blockchain to ensure their integrity and nontampering. The size of each customer's data will also be recorded in the block to set up the reward mechanism.

Now let's move on to the server side, which, after receiving the weights from the clients' nodes, will ensure the generation of the global models. The "Federation" smart contract will also ensure the hashing of global models and their registration on the blockchain.

NFT "Cookies": When Web3 Tech Meets Web2.0 Legacy

As metaverse applications gain more mainstream adoptions down the road, the debate started with Web2 on third-party cookies, dystopian society, and surveillance economy will be ongoing topics of debate. We start with third-party cookies taking new form in NFT marketplaces in this section, and in the following section we will discuss the risks of surveillance economy and dystopian society.

Third-party cookies are cookies that are set by a website other than the one you are currently on. For example, the "Like" button at the Facebook website could store a cookie on your computer, and that cookie can later be accessed by Facebook to identify visitors and see which websites they visited. This allows Facebook to "follow" you to see any website that you have visited. It is like having Facebook put a GPS tracking device in your body and follow you in the virtual world. If we allow this kind of third-party cookie in the metaverse, that would be a major privacy concern, because the metaverse is meant to have us "immersed" into it.

Similar to the third-party cookie problem, in the NFT ecosystem, the issue has been demonstrated in OpenSea (one of the biggest NFT marketplaces) and MetaMask (one of the most popular crypto wallets). This has to do with ERC1155 and ERC721 NFT standard, which allows the creators to set the URL of NFT metadata to the creator's or seller's website. The seller's website can then track the user's IP address. Of course, websites like OpenSea often collect and store visitors' IP addresses in virtue of how the sites function. OpenSea itself likely collects the IP addresses of visitors, like plenty of other sites, apps, or services. But here, an outside third party – an NFT seller or creator – can also gather information on the people viewing the NFT, potentially without their knowledge. (See **Box: IP Address Leak at OpenSea and MetaMask.**)

IP Address Leak at OpenSea and MetaMask

In January 2022, Alex Lupascu, co-founder of privacy and blockchain company Omnia, described how his team discovered that popular cryptocurrency wallet MetaMask had an issue where an attacker could mint an NFT and then send it to a victim to obtain his or her IP address. For a background, a usual lifecycle and interaction with an NFT is as follows:

1. Upload collectible's image on a server.
2. Mint the NFT on the blockchain and only store the holder's address and URL of remote image.
3. Optionally, transfer the NFT to another blockchain address.
4. Holder's crypto wallet reads the blockchain to scan what collectibles it owns and finds the aforementioned NFT.
5. Crypto wallet fetches the remote image from the URL associated with the NFT.

In Alex Lupascu's demonstration, the token directed the user's wallet to a server that grabbed the image to display in the wallet. Because NFTs usually only contain a URL pointing to a server that holds the actual image, rather than the image itself, Lupascu devised a setup where an attacker controlled this server and harvested the user's IP address when the wallet fetched the image. According to Lupascu, the hacker could steal the user's IP address, and in theory the IP addresses can be used to launch a distributed denial of service attack that overloads a specific URL with traffic.

Put it simply, if the server hosting the image is controlled by a malicious actor, then the IP address of the NFT holder is leaked by his mobile phone when the crypto wallet fetches the collectible's remote image. After this demonstration, MetaMask founder Daniel Finlay later said they were starting work to fix the issue raised by Lupascu.

We can view this issue as Web3 technology (NFT and associated smart contract) meeting Web2 (the NFT URL content hosted in Web2) and creating new forms of privacy concerns. There will be ongoing privacy concerns from existing third-party cookies from Web2 world, compounded by new privacy concerns exposed by NFT.

To move beyond Web2 third-party cookie issues, Google announced in March 2021 that it would stop using cookies on its Chrome browser by 2022. (Later the deadline was extended to 2023. And in 2019, Mozilla's Firefox browser started blocking third-party cookies by default.) This doesn't mean that advertisers won't have tools to target you on the country's most popular browsers. Google, in fact, is already testing alternatives to third-party cookies.

Google has created something called "Federated Learning of Cohorts," or FLoC proposal. This, Google says, is about finding a third-party cookie alternative that protects user privacy. The FLoC system, which is pronounced like the word *flock*, would put people into groups based on similar browsing behaviors. This means that advertisers would use only cohort IDs and not individual user IDs to target them. Web histories of users would be kept on the Chrome browser, but Chrome would only provide advertisers with information on a cohort that is made up of thousands of individual web surfers. (This concept is similar to the earlier Federated Learning discussion at **Figure 7.6**.)

One cohort might include thousands of users who have browsed alternative music sites. Others might contain users who are interested in comics or animation. This, Google says, provides advertisers with a powerful tool while protecting the privacy of individual Chrome users. In short, FLoC is a form of interest-based tracking that identifies you based on your "cohort," or a group of people that share similar interests. It seems that FLoC can make it easier for advertisers to identify you.

As Google started developing FloC, governments have enacted legislation to create civil and criminal penalties for companies, marketers, and others who fail to inform consumers that their websites are using cookies. Such legislation includes the General Data Protection Regulation (or GDPR) in Europe, which regulates how personal information is collected, stored, and eliminated. It also includes the California Consumer Privacy Act, or CCPA, designed to protect the privacy of California consumers. Virginia also recently enacted the Virginia Consumer Data Protection Act. In addition, the Washington Privacy Act and the Illinois Biometric Information Privacy Act are both getting close to being enacted.

Under regulatory pressure and privacy advocacy groups' criticism, Google has paused the "FLoC proposal" testing. In early 2022, the search giant proposed a new approach called "Topics," which enables advertisers to place ads based on a limited number of topics that are determined by users' browser activities. The Topics' application programming interface (API) uses the Chrome browser to determine a list of up to five topics – such as "books and literature" or "team sports" – that a user is likely interested in, based on the websites they visit. The features of "Topics" include:

- Topics are determined on a weekly basis.
- They exclude sensitive categories, such as gender or race.
- Users can review and remove topics from their lists, or they can turn off the entire Topics API.
- Topics are kept for only three weeks.

- When the user visits a site that supports the Topics API for ad purposes, the browser will share three topics the person is interested in – each is selected randomly from the user's top five topics in the past three weeks. The site can then share this information with its advertising partners to determine which ads to present.

This approach got some pushback from marketers since "Topic" does not give target needs of consumers. For example, the Topic can be home appliance, which is too big a category for the marketer to place an effective ad on the user. But for privacy concerned advocates, the "Topic" may still contain too much metadata, which can be used to violate users' privacy. Therefore, the alternative solutions to third-party cookies will still be an ongoing debate.

For NFT, the service provider such as OpenSea, Metamask, and other NFT marketplaces and wallet providers, can mitigate the privacy issue by downloading the URL content from the servers controlled by them, instead of letting individual users download directly from an NFT creator's website. This way, the user only interacts with NFT service providers, not directly with NFT creators for URL content downloading. Of course, if a user uses Chrome to visit OpenSea or MetaMask, the Web2 third-party cookie tracking is still a privacy concern – until Google comes out with a good solution.

Surveillance Economy and Dystopian Society

MEV and Surveillance Economy

Data has commercial value. Because data has become a critical resource in the new digital economy, the internet giants are more often *proactively* collecting user data. (See **Box: Buying Faces.**) Furthermore, they are collecting every aspect of user data, whether identity data, network data, or behavioral data. Because the legal development lags the age of Big Data, network service providers globally have all taken liberties with the collection and use of personal information in mobile apps.

For example, a photo-editing app should have no right to demand user location, ID number, or fingerprints. And why should a weather forecasting app ask for access to a user's contact list? As our interactions with companies and their applications move from screens in our hands to headsets on our faces, the potential for invasive data collection grows. That's the risk of surveillance economy in the metaverse.

According to Wikipedia:

Surveillance capitalism is an economic system centered around the commodification of personal data with the core purpose of profit-making. The concept of surveillance capitalism, as described by Shoshana Zuboff, arose as advertising companies, led by Google's AdWords, saw the possibilities of using personal data to target consumers more precisely. Increased data collection may have various advantages for individuals and society such as self-optimization, societal optimizations (such as by smart cities), and optimized services (including various web applications). However, collecting and processing data in the context of capitalism's core profit-making motive might present a danger to human liberty, autonomy and wellbeing.

Buying Faces

In China's rural areas, such as Henan province in the middle of the country's heartland, the peasants of small villages found they can trade images of their faces for daily supplies. Villagers were asked to stand in front of a camera and slowly rotate side to side to have their face pictures taken. These face pictures would be used in AI software to distinguish between real facial features and still images. In return, the peasants were given kettles, pots, and teacups.

"Buying faces" is not a China-only phenomenon. In the summer of 2019, Google announced a face-unlock feature for its upcoming Pixel 4 phone, which claims to be just as accurate and fast as the iPhone's Face ID. In various cities across the country, Google employees were reportedly doling out $5 Starbucks and Amazon gift cards to people on the street in exchange for a facial scan. Google confirmed that such "field research" was an effort to collect diverse face-scanning data to ultimately improve the accuracy of its upcoming Pixel 4's facial recognition technology.

Metaverse builders may repeat the same mistakes that Web2.0 companies have made if privacy is not the top priority in the design of metaverse applications. With Web3, the central design element is the decentralized nature of data and user's self-control of data. Nevertheless, the on-chain transactions are still recorded on the distributed ledger. Companies such as Chainalysis, Dune Analytics, and CoinDesk have the resources and talent to perform on-chain and off-chain analysis, and may gain additional information about individuals.

For example, crypto miners can look into the mempool for transactions to obtain miner extractable value (or **MEV**). MEV is a measure of profit that blockchain miners can make through their ability to arbitrarily include, exclude, or reorder transactions. Because miners (and validators) process and validate transactions, they have full visibility of all transactions, and consequently they're able to reorganize the transaction at their own will and front-run other users' transactions. This way, such miners can make profits at the cost of other users.

The privacy-preserving computing technologies introduced in this chapter can help protect data privacy and minimize the MEV issues that currently exist in almost all major public chains. More research on privacy-preserving computing and storage, as well as token economy, is needed to avoid the same mistakes of Web2 technologies where data can be used to disrupt the democratic process (e.g., Cambridge Analytica election interference using Facebook user data), or data leaks that impact millions of individuals (e.g., at Yahoo, Google, Facebook, LinkedIn, and Equifax).

Dystopian Society

Imagine a society where you live under the rules of some organization controlled by a tiny group of privileged elites. It may be a despotic government, a religious organization, an all-powerful global corporation, or a DAO created by a metaverse community. This organization controls many aspects of your life. You are told what to think and how to act through propaganda and brainwashing. Individual thoughts and actions in contradiction to what is permitted are not tolerated and are severely punished if discovered by the authorities. Your privacy is not protected and is controlled by the organization. This is but one of many visions of a dystopian society.

Dystopian societies are often depicted in science fiction literature and film. Examples of dystopian fiction include George Orwell's *1984*, Anthony Burgess's *A Clockwork Orange*, Aldous Huxley's *Brave New World*, and Margaret Atwood's *The Handmaid's Tale*. Well-known dystopian films include the recent *Hunger Games*, *Children of Me*. These novels and movies have served as warnings about a dystopian future of technology gone wrong.

The following are some of the scenarios that the Metaverse may create in a dystopian society nightmare:

- Big Tech companies build metaverse applications to amass more data, and then use advanced AI/ML to gain more insights about the users' behaviors and privacy. The goal of Big Tech is to maximize profits, which may be at the expense of user privacy and user freedom. Moreover, the large tech companies can impose their own dystopian rules or collaborate with oppressive governments to suppress individual freedom and privacy.
- The Web3 community can build a DAO to attract global metaverse users and the DAO can be controlled by a few "crypto whales" who have the majority of voting power or stakes in the DAO. The whales can impose dystopian rules out of their own interests at the expense of global users who have very little voting power.

The collaboration of Big Tech, Web3 DAO community, and the government may collude to create an all-powerful metaverse ecosystem that may impose dystopian rules and breach individual privacy and even basic human rights.

In summary, there are many thorny data privacy questions we would have to answer, as the vision for a sophisticated, converged metaverse becomes a reality. Simply put, the metaverse blurs the lines between the real and the virtual at a scale never seen before. As a result, data privacy and security are major concerns for metaverse companies, developers, and users alike. While we still have much to understand about how metaverses will take shape, we have enough of a framework for thinking about data privacy and data security (to be discussed in the following chapter) to ensure that we do not make the same mistakes we allowed in our current internet.

CHAPTER 8

Metaverse Security

- Blockchain and Metaverse: Marriage in Heaven?
- Identity in Metaverse: Wild Wild West
- Metaverse Data Security: Chronic Pain
- Smart Contract Security: Maybe Not So Smart
- Ransomware Attack in Metaverse: Is it Possible?
- Supply Chain Software Attack: A Real Danger?
- Quantum Computing: Future Challenges and Opportunities
- Extended Reality (XR): Novel Security Risks

Blockchain and Metaverse: Marriage in Heaven?

Blockchain's encryption, immutability, and decentralization attributes make it a great choice for securing data. Blockchain-enabled data security methods help ensure confidentiality, integrity, and availability of information (see **Figure 8.1**).

First, **confidentiality.** The asymmetric encryption mechanism used in the public key provides pseudo anonymity of the user. Privacy coin leveraging zero knowledge, multipart secure computing, ring signature, coin join, and threshold signatures provides enhanced confidentiality of blockchain transactions. The layer two technology in Ethereum, the lightning

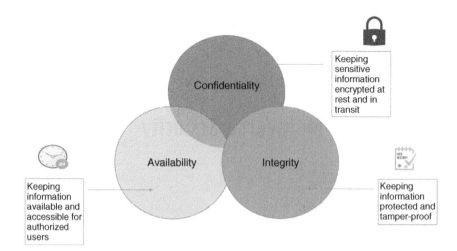

Figure 8.1   The CIA of Blockchain

network in Bitcoin, and channels in HyperLedger Fabric are also excellent mechanisms for confidentiality in addition to performance gains.

For example, Hyperledger Fabric can be defined as a logical entity that represents a grouping of two or more blockchain network members/participants for the purpose of conducting private and confidential transactions between themselves. Its *channels* are private and permissioned and meet the following requirements to enhance the confidentiality of the network:

- Not anyone and everyone can join a channel.
- Each peer joining a channel has its own identity.
- This identity is given by a membership services provider (MSP).
- Each peer in a channel must be authenticated and authorized to conduct transactions on that channel.

Second, **integrity.** Blockchain data or transaction has to be signed, consensus agreed, and appended in the ledger. The ledger is globally distributed for and the integrity is ensured at global scale, not like a centralized database.

Third, **availability.** Data on blockchain is hosted redundantly by many network nodes; the availability of data (if not considering the response time in data query) can be ensured because nodes are distributed and controlled with a good incentive mechanism by network participants. In addition, Smart contract, and protocol for data (in terms of ownership, pricing, and exchange) encourage and incentivize availability and sharing of data.

Although blockchain technology introduced new challenges to cybersecurity (we have to think about the cybersecurity of blockchain, too), the blockchain security features can be leveraged by metaverse applications for data confidentiality, data integrity, and data availability. It certainly seems like a marriage made in heaven if blockchain can be used properly in metaverse applications. However, as this chapter will discuss, the data security issues in the Web3 remain more complex than the marriage itself.

Identity in Metaverse: Wild Wild West?

In the Metaverse, users' digital identity is where intruders strike first. The metaverse is expected to bring a shift from login IDs and usernames to enhanced digital avatars. Your metaverse profile may contain much more personal information than your current Google or Facebook account. If you are not careful, it may integrate your entire digital life and your personality — not only your unique online (and offline) identity, but also your bank account and other sensitive data. Protecting digital identities against theft will be a critical factor for metaverse applications; equally important, the community must ensure that metaverse users cannot fake their identities.

In history, fraudsters have claimed to be deposed princes with fortunes to share, or sweepstakes hosts desperately trying to reach you. When the internet developed, these schemes refranchised digitally by email, text messaging, and social network communication. Playing this forward, identify theft will be further upgraded in the metaverse. It won't be a fake email from your bank. It could be an avatar of a teller in a virtual bank lobby asking for your personal identification. It could be an impersonation of your social network friends inviting you to a multiplayer game in a malicious virtual room.

Therefore, solving for identity in the metaverse is a top concern. With a paradigm shift comes a great responsibility – but who will be the gatekeeper, and how can you ensure that your information is kept safe? The security of this new environment will lie on the shoulders of the companies involved with the metaverse as they will have to assume the role of identity verifiers. They will have to find a way to authenticate identity, protect the identity, and defend against misuse and fraudulent use of identity.

However, organizations need to know that adopting metaverse-enabled apps and experiences won't upend their identity and access control. There are many open questions about identity protection, and they are interesting challenges for the metaverse community as well as the cybersecurity providers. For example:

1. **Will the metaverse make the identity crisis worse in industries like gaming, where attempts of hacking, tampering, cheating, and theft are already prominent?** For example, "deepfakes" is one challenge. Deepfake uses a deep-learning system to produce persuasive counterfeits by studying photographs and videos of a target person from multiple angles, and then mimicking its behavior and speech patterns. Once a preliminary fake has been produced, a method known as GANs, or generative adversarial networks, makes it more believable. The GANs process seeks to detect flaws in the forgery and then improve the fake by addressing the flaws. There are many deepfakes as service providers in the market that can simulate actual identity presentation via voice, image, and video data. When more and more personal

data become exposed in 3D, immersive gaming (see related data privacy discussion in **Chapter 7**), we may see explosive growth of deepfakes.

2. **The ambiguity of ownership in the metaverse will create a lot of intellectual property disputes.** Who will truly own in-game contents or NFT items, the publishers, or users? Who drives the sales of the contents and who represents the users who generate content? The linkage between digital assets and individual identities is a huge topic for the metaverse.

3. **The metaverse ecosystem will process enormous amounts of personal data.** This will be subject to increasingly stringent data regulations as the existing social networks face. This may be a heavy compliance burden for tech start-ups, especially for smaller companies keen to build up the metaverse. We must make identity manageable for enterprises in this new world.

Again, take "deepfakes" as an example. In January 2022, the Chinese regulator proposed draft rules to crack down on the spread of "deepfakes" at several platforms. As proposed by the draft rule, "Where a deep synthesis service provider provides significant editing functions for biometric information such as face and human voice, it shall prompt the (provider) to notify and obtain the individual consent of the subject whose personal information is being edited." Those first-time violators of this consent rule will be fined 10,000 yuan ($1,600) to 100,000 yuan ($16,000).

What's encouraging is that many researchers and standard organizations such as W3C and Decentralized Identity Foundation have been working hard to define the security requirements of decentralized identity. The newly formed organization named "Trust over IP" is also trying to provide a robust, common standard and complete architecture for internet-scale digital trust, which can be used by metaverse applications. Constructive steps also include making things like multifactor authentication (MFA) and passwordless authentication integral to platforms.

As illustrated by the identity security issue, any metaverse application faces two basic sets of security problems: familiar challenges technologists have been dealing with for decades, and brand new ones built specifically for metaverse settings. Tightening user identity security is the first step of metaverse data security efforts, and in the following section we will cover more complex issues.

Metaverse Data Security: Chronic Pain

It is not clear at this point that the newly branded Facebook company, Meta, will leverage any of the Web3.0 technologies. The key tenet of Web3.0 is self-sovereign identity, meaning that the user of the platform will own the data with selective consent. The Metaverse will generate tons of new data, a few magnitudes more than in the past. If a centralized entity such as Meta or LinkedIn is holding this data, then past data leaks on these platforms will repeat themselves and will be so in a more destructive fashion. Here are some examples of data leaks in Big Tech companies in 2021.

In April 2021, half a billion Facebook users' information was leaked. Details in some cases included full name, location, birthday, email addresses, phone number, and relationship status. The data could be used for carrying out social engineering attacks such as phishing. Typically, a social engineering attack involves a bad actor imitating a legitimate person or organization, including a bank, company, or coworker, in order to steal data such as login credentials, credit card numbers, social security numbers, and other sensitive information.

In June 2021, data associated with 700 million LinkedIn users was posted for sale in a Dark Web forum. This exposure impacted 92 percent of the total LinkedIn user base of 756 million users. The leaked data included the following: email addresses, full names, phone numbers, geolocation records,

LinkedIn username and profile URLs, personal and professional experience, genders, other social media accounts, and other details. The hacker scraped the data by exploiting LinkedIn's API. LinkedIn claims that, because personal information was not compromised, this event was not a "data breach but, rather, just a violation of their terms of service through prohibited data scraping." But the leaked data was sufficient to launch a deluge of cyberattacks targeting exposed users. In fact, most cybersecurity professionals classified the incident as a data breach.

In March 2021, Microsoft suffered a significant security breach. It involved more than 250,000 victims across 30,000 organizations worldwide, when its on-premises exchange servers suffered four zero-day exploits. The attack on Microsoft made many headlines as it released user login credentials on the affected servers and gave admin privileges to the attackers. Microsoft stated that a new strain of ransomware infiltrated its server, encrypting all data and rendering them unusable while demanding money for updates. After the attack, Microsoft claimed that it had fixed more than 90 percent of the servers.

As Big Techs embrace metaverse technologies and business models for the next-generation internet, this kind of attack will pose more risks to individual users or organizations using the platform. The data collected from IoT devices, AR/VR headsets or other wearables, virtual working spaces, chats, shopping histories, cryptocurrency transactions, games, and NFT transactions can be stored centrally in BigIT's cloud environments. The data stored on the cloud or BigIT data center can be a sweet target by hackers, and there will certainly be future headline news about massive data leaks.

What is the solution? Of course, the traditional defense in-depth approach with zero trust model can help to mitigate metaverse security risks (see related discussion in **Chapter 7**). Nevertheless, the centralized data ownership by Big Tech companies remains the key problem from security, privacy, and fairness perspectives. A better solution is to return the ownership and storage of data back to individual users of the platform. This is a central idea of Web3. This includes self-sovereign identity (SSI) and user consent-based data ownership.

Before we dive into the concept of SSI and data ownership, we need to understand the concept of public key cryptography and how they are used in blockchain.

Public Key Cryptography

Public key cryptography uses a public key and a private key to perform different tasks. Public keys are widely distributed, while private keys are kept secret.

Using a person's public key, it is possible to encrypt a message so that only the person with the private key can decrypt and read it. Using a private key, a digital signature can be created so that anyone with the corresponding public key can verify that the message was created by the owner of the private key and was not modified since.

Blockchain makes extensive use of public key cryptography. Major cryptocurrency companies like Bitcoin, Ethereum, and Bitcoin Cash function using three fundamental pieces of information: the address, associated with a balance and used for sending and receiving funds, and the address's corresponding public and private keys. The generation of a bitcoin address begins with the generation of a private key. From there, its corresponding public key can be derived using a known algorithm. The address, which can then be used in transactions, is a shorter, representative form of the public key.

The private key is what grants a cryptocurrency user the ownership of the funds on a given address. The blockchain wallet automatically generates and stores private keys for you. When you send from a blockchain wallet, the software signs the transaction with your private key (without actually disclosing it), which indicates to the entire network that you have the authority to transfer the funds on the address you're sending from.

The security of this system comes from the one-way street that is getting from the private key to the public address. It is not possible to derive the public key from the address; likewise, it is impossible to derive the private key from the public key. In most crypto wallets, there is a 12-word or 24-word mnemonic that is a seed for a private key. Public key cryptography is the foundation for self-sovereign identity and data ownership.

Self-Sovereign Identity (SSI) and Data Ownership

What Is SSI?

Self-sovereign identity systems use blockchains – distributed ledgers – so that decentralized identifiers can be looked up without involving a central directory. Blockchains do not solve the identity problem by themselves, but they do provide a missing link that allows things we have known about cryptography (but cannot apply) for decades to become usable. Blockchain technology allows people to prove things about themselves using decentralized, verifiable credentials as they do offline (see **Figure 8.2**).

Who Is an Issuer?

An issuer is an entity that issues a credential. For example, a test management facility like a hospital that issues a patient record (e.g., Covid-19 vaccination status). Issuer has the right to revoke a credential.

Who Is a Holder?

A holder is an entity that has lifecycle control over the issued credentials like sharing and deleting. For example, a patient could hold a credential issued by an issuer on his or her wallet (a wallet could be an app that stores users credential data locally or a custodial wallet managed on behalf of a holder).

Who Is a Verifier?

A verifier is an entity that verifies if the credential shared by a holder is valid (i.e., if the credential comes from a trusted

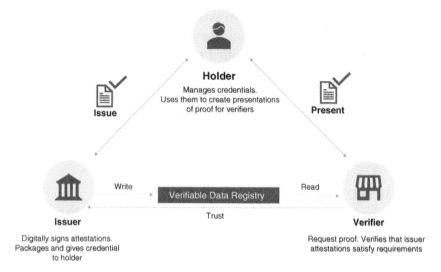

Figure 8.2   Self-Sovereign Identity (SSI)
Source: Affinidi

issuer, not revoked by the issuer). For example: An access management system installed at a facility like airport that allows / denies access based on whether the holder has completed Covid-19 vaccination. Verification could be a combination business logic like "is the credential issued in the last 14 days" and "is it issued by an issuer that is recognized."

What Is a Verifiable Credential (VC)?

The term *credential* can imply any (tamper-resistant) set of information that some authority claims to be true about you, and that enables you to convince others (who trust that authority) of these truths. For example, a diploma issued by a university proves you have an educational degree. A passport issued by a government of a country proves you are a citizen.

Every credential contains a set of claims about the subject of the credential – that is, about the holder. These claims are made by an issuer. To qualify as a credential, the claims must be verifiable in some way. This means a verifier must be able to determine the following:

- Who issued the credential
- That the credential has not been tampered with since it was issued
- That it has not expired or been revoked

With physical credentials, this is accomplished through some proof of authenticity embedded directly in the credential itself like a chip or hologram. It can also be done by checking directly with the issuer that the credential is valid, accurate, and current. But this manual verification process can be difficult and time-consuming – a major reason why there is a worldwide black market in falsified credentials.

This brings us to one of the fundamental advantages of verifiable credentials: using cryptography and the blockchain, they can be digitally verified in seconds. This verification process can answer the following three questions:

- Is the credential in a standard format, and does it contain the data the verifier needs?
- Is the credential still valid – that is, not expired or revoked?
- If applicable, does the credential (or its signature) provide cryptographic proof that the holder of the credential is the subject of the credential.

SSI will allow users or holders in **Figure 8.2** to maintain and have the ownership of its own digital identity across multiple platforms while selecting the information they wish to share on each. This mode of interaction would drastically transform the current digital marketplace that has turned personal data into a commodity. Identity is going to be returned, through blockchain, back to the individual so that the individual will own their data and then be able to marshal it out based on what's best for them, as opposed to how Facebook or LinkedIn or other people may want to exploit it.

With SSI, the security focus can be changed from centralized platform to wallet security, SSI security, and Access Control using smart contracts for data access. One such example is OpenZeppelin, which provides smart contracts for secure data access. The basic building block of SSI is public key cryptography. The owners of an SSI use a digital wallet with a private key to hold their digital identity, and they use this identity in conducting data ownership verification, data related transactions, and metaverse data exchanges. Users can encrypt the data and store the data in an IPFS file storage or NFT storage space, and the ownership of the data can be verified using the user's private key to digitally sign the data. The data can be encrypted by using the public key, so only the user can decrypt the data using the corresponding private key.

In practice, an encrypt key can be derived from the private key. This encrypt key can then be used as a symmetric key to both encrypt and decrypt the data for higher performance in data encryption, since encryption using public key cryptography usually is still very slow, especially with a large amount of data. SSI is a European Union standard. The European Union is creating an eIDAS compatible European Self-Sovereign Identity Framework (ESSIF). The ESSIF makes use of decentralized identifiers (DIDs) and the European Blockchain Services Infrastructure (EBSI), and (eIDAS stands for electronic Identification, Authentication and Trust Services defined by the EU).

The World Wide Web Consortium (W3C)'s Credentials Community Group (CCG), in collaboration with Decentralized Identity Foundation (DIF), is actively working on building a suite of DID) standards that cover core DID attributes, DID authentication and discovery, verifiable claims, DID secure communication, secure data storage, and wallet security. These suites of standards will clarify how identity and data ownership are defined in the metaverse platform. Of course, its most basic component is still the private and public key pairs used in public key cryptography. The protection of private keys will become a very important defense measure for metaverse data security.

Data Leak Cases in NFT Projects

If a hacker steals your private key or the mnemonics used for the private key, you will lose your data and assets associated with this key. Worse, if you are a metaverse company that hosts users' funds, and a hacker is able to hack the server with the private key, then the financial loss can be huge.

This was what happened to a $140 million hack from a Polygon gaming platform named Vulcan Forged (NFT Marketplace) in December 2021. According to company CEO Jamie Thomson, the hacker was able to attack the semi-custodial wallets that Vulcan Forged helped manage for its customers. The problem wasn't with its wallet solution provider, Venly, but a vulnerability with Vulcan Forged. "What's happened is someone's exploited our servers, got the Venly credentials, and used it to extract the private keys of the MyForge users," Thomson said in a video shared on the company's social media accounts after the attack. (MyForge is an asset management tool that displays users' crypto and NFT holdings.) "Going forward, of course, we're going to be using nothing but decentralized wallets so we never have to encounter this problem again."

Another example was the hack of a Hong Kong–based NFT project in December 2021. In that case, the hacker stole an administrator account of the project's group chat on Discord, a popular online instant messaging service. Launched on November 27, 2021, Monkey Kingdom, which is comprised of 2,222 digital portraits of the mythical hero Monkey King dressed in different styles, has quickly become one of the most talked-about NFT projects in Asia, with endorsements from celebrities including Steve Aoki, JJ Lin, and Ian Chan of the Hong Kong–based boy band Mirror.

The hacker posted a "phishing" link in the group chat, just as the project kicked off a new sale in earnest. Buyers lost more than 7,000 solana, a popular cryptocurrency, to the scam, which amounted to nearly US$1.3 million. Phishing is a common form of online fraud used to steal user data, including login credentials and credit card numbers. It occurs when an attacker, masquerading as a trusted entity, dupes a victim into opening an email, instant message, or text message. It is now being used to breach access to users' cryptocurrency wallets. This example demonstrates that if a hacker gets the administrator account of a Web2.0 application, he can "fish" the users to send him the funds even if the user has full control of their private key.

Therefore, self-ownership of data and identity is only the first step for data protection in the metaverse. Traditional security practices, such as defense in depth and zero trust approach, are still needed. In addition, users must know how to protect the private keys used for transactions in metaverse. Furthermore, the protection from private keys is not enough; wallet security, secure data storage, and secure decentralized identity communication will need to be in place to enhance data protection in the Web3 realm.

For example, one of the W3C and DIF initiatives is to define a common terminology for understanding the security requirements applicable to wallet architectures and wallet-to-wallet and wallet-to-issuer/verifier protocols. The wallet security working group will classify, specify, and describe security architectures common to wallets (such as risks, motivation, etc.) and produce guidelines for how to classify and specify the security capabilities of verifiable-credential wallets such as key management, credential storage, device-binding, credential exchange, and the backup, recovery, and portability of wallets.

The secure data storage working group is tasked to create one or more specifications to establish a foundational layer for secure data storage (including personal data), specifically data models for storage and transport, syntax, data at rest protection, Create, Read, Update, Delete (CRUD) API, access control, synchronization, and at least a minimum viable HTTP-based interface compatible with W3C DIDs/VCs.

The DID secure communication group is working on producing one or more high-quality specs that embody a method ("DIDComm") for secure, private and (where applicable), authenticated message-based communication, where trust is rooted in DIDs and depends on the messages themselves, not on the external properties of the transport(s) used.

Going forward, metaverse data security will be an ongoing issue that will attract researchers and practitioners to work on finding countermeasures to defend metaverse ecosystem participants from falling victim to data breaches.

Smart Contract Security: Maybe Not So Smart

Most metaverse platforms will leverage smart contracts for various business logic such as payment conditions, incentive mechanism, play-to-earn logic, staking rewards, and many innovative business logics. Indeed, a metaverse platform without smart contract and underline base layer blockchain would not really make much sense. However, the main challenge to smart contracts is the security issue.

Smart contracts are at the core of every blockchain, and they exhibit the following characteristics:

- **Distributable.** Smart contracts can be validated by every participant in the network, similar to the regular transactions on a blockchain.
- **Immutable.** By design, smart contracts cannot be changed or tampered with once they are released.
- **Transparency.** All the terms and conditions of the agreement in smart contracts remain visible to network participants.

- **Cost efficiency.** Smart contracts remove the necessity for additional validation of the agreement and the extra expenses.
- **Accuracy.** The terms of the agreement are written in the form of code, and smart contracts follow these terms and conditions without any exceptions.

The terms of the smart contract agreement are written directly into the code. Hence, smart contracts can securely and efficiently transfer the funds or information between the participants without any need for mutual trust. The involvement of regulators and intermediaries is no longer necessary. Therefore, a smart contract needs to be highly secure with no bugs, loopholes, or vulnerabilities hidden in its programming codes.

Smart contract has been the main target of attacks in the past few years, and that trend has accelerated in the last two years (2020 and 2021) with DeFi and NFT applications. Once deployed, the business logic implemented in the smart contract will be permanent unless there is emergence stop or pause logic in place. With metaverse applications, we will see more smart contracts and more attacks on the smart contract. Following are a few case studies relating to hacks in NFT and GameFi due to smart contracts' vulnerabilities.

CryptoPunks Smart Contract V1 Error

Launched in 2017 as the first NFT project, CryptoPunks suffered from a severe smart contract bug that led to the seller of the NFT token not receiving any payment (despite the sales). The bug was found after all the 10,000 punks were traded and the secondary market started listing the token. The smart contract essentially refunded the buyer of NFT tokens after they made payments to buy NFTs, and consequently the seller cannot get any payment for selling NFTs. This bug was later fixed in a new version of the smart contact.

**Creators of CryptoPunks Apologize
For 'V1' Ethereum NFT Sales**

```
function punkNoLongerForSale(uint punkIndex) {
    if (punkIndexToAddress[punkIndex] != msg.sender) throw;
    punksOfferedForSale[punkIndex] = Offer(false, punkIndex, msg.sender, 0, 0x0);
    PunkNoLongerForSale(punkIndex);
}
```

Figure 8.3   Creators of CyptoPunks Apologize for V1 NFT Sales
Source: CryptoBullsClub, EatTheBlock

Figure 8.3 shows the code that caused the problem. The NFT's index was overwritten inside the function and the NFT seller's file is changed to the message sender (the buyer) so all the fees paid to seller are actually transferred back to the buyer.

Hashmasks Smart Contact Bug

A bug was reported by a security researcher named Samczsun in the Hashmasks art sale during the late stages, but luckily, there was no damage and Hashmasks was able to take remedial steps in time. Samczsun raised a flag about a potential bug in the Masks.sol smart contract of Hashmasks relating to the mintNFT function. Had attackers been able to exploit the bug, they would have minted more than 16,384 Hashmasks. Somehow the bug could not be discovered during the testing phase, and Hashmasks awarded Samczun $12,500 USD Coin (USDC, a kind of stablecoin developed by Coinbase and Circle joint venture) for discovering the bug.

Twitter Profile File Picture (PFP) NFT Hack

In January 2022, Twitter announced a feature that enables Twitter Blue subscribers to designate their profile pictures as official NFTs — thus, in theory, irrefutably "proving" their JPEGs are authentic pieces of the likes of CryptoPunks or Bored Apes at a glance. It's the ultimate online flex – or showing off in our status-obsessed digital age. It's also, unfortunately for NFT enthusiasts, extremely easy to fake.

During the same month, a "white hat hacker" was able to get a CryptoPunk look-alike NFT as his official Twitter profile picture – hexagonal and all. The hacker used an old smart contract on the Ethereum mainnet, and simply changed the token URI – the place that associates the tokens with the images and other metadata that make them an NFT – to match the token URI of another collection. In this case, the hacker made his tokens to look like Bored Apes.

Twitter's new feature has put its trust completely in Open-Sea for verification – and this was the SINGLE checkpoint that prevented hackers from making a PFP identical to a "real" ape. Twitter is overly reliant on some small visual signals to show which of these two is from the actual, verified collection. Limiting the feature to "Twitter Blue" subscribers seemingly lends additional credibility. However, given that the hexagon is just as easy to fake as a regular NFT-based profile picture, the feature doesn't work as intended.

The following are the potential mitigation strategies for the issue:

- Always verify the official smart contract address used to mint (or set URI) for the metadata used for NFT collections.
- Increase the visibility of verification signals in the detail view. One of the most valuable aspects of this feature is proof of authentic ownership.
- Stop relying entirely on OpenSea for this metadata. If OpenSea goes down, so does the feature. This also makes

OpenSea an official arbiter of whose collections are credible on Twitter — do we really want to designate OpenSea as the central authority on authenticity?

- OpenSea should prohibit the use of identical collection photos and word-for-word, copy-paste descriptions for collections.
- Decentralize the verification of collections. As more platforms begin to support Web3 features – for example, verified ownership of a profile picture from an elite collection – they'll need to be able to tell the difference between legitimate collections and fake ones. Today, this is still left up to centralized entities like OpenSea. Social media platforms will either rely on third parties like OpenSea for verification, or they'll verify them on their own.

A better solution is a decentralized way to legitimize collections – one that acts as the single source of truth (a Chainlink-like Oracle solution for NFT may be a solution) that every platform can check against. This would also help lesser-known (but very legitimate) collections get verified, too.

Following are some security design patterns for Solidity smart contracts, which is the most used smart contract for metaverse applications:

- **Mark untrusted contracts.** It is essential to specify your variables, methods, and contract interfaces during any interaction with external contracts. It applies to the functions that call external contracts.
- **Prevent state changes after external calls.** When using raw calls or contract calls, there is a possibility that malicious code may get executed. Although the external contract is not malicious, the malicious code may undergo execution by any contract it calls. The malicious code can hijack the control flow and result in causing vulnerabilities due to reentrancy. Therefore, while making a

call to an untrusted external contract, prevent the state changes right after the call. This pattern is known as the check-effects-interaction pattern.

- **Error-handling in external calls.** The low-level call methods in solidity, which work on raw addresses, never throw an exception but return to a false value when an exception is encountered. On the contrary, the contract calls result in propagating a throw automatically on discovering any throw function like doSomething(). So, when you prefer to choose low-level call methods, ensure you handle the possibility of call failure by monitoring the return value.
- **Prefer pull over push for external calls.** The external calls are prone to accidental failure. It is generally applicable in payments, where the users can withdraw or pull the funds automatically instead of pushing funds. It also minimizes the issues associated with the gas limit.

  It is better to isolate each external call into its transaction, which the call recipient initiates.
- **Manage the function code: conditions, actions, and interactions.** A good practice to ensure smart contracts' security comes up with a better structure of all functions. It involves checking all preconditions, making changes to the states, and dealing with other smart contracts.

Many other smart contract security design patterns have been used or developed by smart contract firms such as OpenZeppelin, CertiK, Chainlink, and KnownSec. It is important for Metaverse application developers to follow the security design patterns when writing smart contracts to avoid loss of funds and to protect users' digital assets.

The bottom line is that a smart contract is neither smart nor a contract. It is not smart because it is not an AI or ML algorithm, and it can cause security issues. It is not a contract, since most jurisdictions do not have laws governing smart contracts. Special care and security auditing must be in place to defend against the hackers prying on funds deposited into smart contracts.

Ransomware Attack in Metaverse: Is It Possible?

What is ransomware? Ransomware or ransom malware may be a form of malicious software package that stops users from accessing their system or personal files and demands a ransom payment to regain access. Cyber-criminals sometimes target high-profile people, companies, and even governmental establishments.

As illustrated in **Figure 8.4**, ransomware locks a victim's laptop or networked server through encrypting data and/or stealing sensitive data and threatens to sell or reveal the data if the victims do not pay a ransom. The ransom (for the decrypting key) is sometimes in cryptocurrency type – most frequently bitcoin, because bitcoin (and other cryptocurrencies) allows cybercriminals to receive funds with a high degree of anonymity, making transactions difficult to track. Failure to comply with the hacker's demand can result in a permanent loss of the info.

Ransomware propagates through malicious email attachments, infected apps, infected storage devices, and compromised websites. There have been cases where attackers used remote desktop protocol and alternative approaches that do not require any kind of user interaction. Over the past

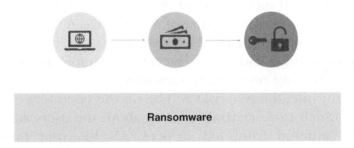

Figure 8.4   Illustration of Ransomware

decade, ransomware has become one of the most prolific criminal business models in the world. (See **Box: Colonial Pipeline Ransom.**)

Colonial Pipeline Ransom

Of all the cyber and ransomware attacks in 2021, the breach of Colonial Pipeline in late April had the most news coverage. As most Americans are directly impacted by gasoline shortages, this attack hit close to home for many consumers. The DarkSide gang was behind the attack and targeted the firm's billing system and internal business network, leading to widespread shortages in multiple states. To avoid further disruption, Colonial Pipeline eventually gave in to the demands and paid the group $4.4 million in bitcoin.

This attack was particularly dangerous because consumers started to panic and ignored safety precautions. Some East Coast residents tried to hoard gasoline in flammable plastic bags and bins, and one car even caught on fire. After the chaos receded, government officials confirmed that Colonial Pipeline's cybersecurity measures were not up to par, and the attack might have been prevented if stronger protection had been in place. Thankfully, US law enforcement was able to recover much of the $4.4 million ransom payment. The FBI was able to trace the money by monitoring cryptocurrency movement and digital wallets. But finding the actual hackers behind the attack will prove a lot harder.

In Metaverse, the "attack surface" for ransomware will be even bigger due to the various new technologies and the huge amount of data accumulated. Thus, the ransomware risk is much harder to manage if the builders and operators of a metaverse application do not adopt defense in depth (did) approaches and monitor security risks of emerging technologies. The decentralized nature of metaverse can mitigate some level of risks since individuals, not the platforms, have the ownership of the data and assets. Hackers will simply steal the data or assets, without resorting to ransomware.

Meanwhile, the real risks will be on the operators of metaverses, which store centralized data about the users and businesses gathered from IoT devices, AR/VR, user browsing histories, and user transactional data. It is critical to seek some mitigation strategy, and blockchain is a potential solution for ransomware.

Blockchain is characterized by immutability and integrity. If a malicious actor attempts to alter the data, every change will be immediately noticed by the system and every other network participant. These design choices make blockchain ideal for data storage because it is an append-only structure, which means that data can only be introduced into the system, and it can never be completely deleted. Any changes made are stored further down the chain, but the network node can always see when the changes occurred, who made them, as well as the previous version of the data.

It is safe to assume that a blockchain-powered database can be an ideal solution to ransomware or other types of data hijacking. With blockchain, metaverse platforms can let platform participants become true owners of their own data. The creation of portable user-owned data means that each participant can choose who has access to their data, move to another metaverse platform without the risk of losing any data, and give instant access to an authorized party with consent. At the same time, metaverse platforms can benefit from an increase in data interoperability, integrity, and security.

In the case of Colonial Pipeline (and many similar ransom cases), a blockchain-powered system would have completely recovered the damages by simply restoring the data records to their previous versions. Therefore, even though ransomware may still show up in metaverse applications, especially when Web2.0 technologies are still used and Big Techs are not giving up the control and storage of user data, we could mitigate the risks caused by ransomware with blockchain technology,

Supply Chain Software Risks: A Real Danger?

What is a supply chain software risk? In a simplified definition, supply chain software attack risk occurs when your application uses a third-party software as building blocks in your application and that third-party software has been compromised and embedded with malicious code, which can then be used to impact your application and introduce risks to your application. In short, threat actors may turn legitimate software into a weapon.

Software supply chain attacks grew by more than 300 percent in 2021, according to a study from Argon Security, recently acquired by Aqua Security. The report found that the level of security across software development environments remains low, and every company evaluated had vulnerabilities and misconfigurations that could expose them to supply chain attacks. Among the attacks, the SolarWinds breach is one of the most prominent examples of software supply chain attacks. (See **Box: SolarWinds Supply Chain Attack.**)

SolarWinds Breach

SolarWinds is a prominent software company that provides thousands of organizations worldwide with numerous technical services and system management tools for infrastructure and network monitoring. The breach occurred through the company's IT performance monitoring system called Orion in 2021. Through this hack, the threat actors gained access to the systems, data, and networks of thousands of SolarWinds customers who were using the Orion network management system for managing their IT resources.

The hackers inserted malicious code into the Orion network management system, which was used by numerous government agencies and multinational companies globally. Due to the addition of this malicious code, the SolarWinds Orion Platform created a backdoor that allowed the hackers to access accounts and impersonate users of victim organizations.

The malware could access system files and seamlessly blend in with legitimate SolarWinds activity without being detected. The hackers installed this malicious code into a new batch of software, which was then sent out to customers by SolarWinds as an update at the beginning of March 2020. More than 18,000 customers of the company installed the update, allowing the malware to spread undetected. The hackers used this hidden code to access the IT systems of SolarWinds customers, using them to install even more malware.

Multiple government agencies and commercial industry verticals around the world were affected by the infamous SolarWinds hack. According to an SEC filing by SolarWinds, around 18,000 of its customers were using the vulnerable versions of the Orion platform. Even several government departments in the US such as Homeland Security, Commerce, State, and Treasury were affected by this breach. A reputable cybersecurity company, FireEye, is the first known victim of this breach and was also responsible for disclosing the attack in December 2020. Many other NGOs and Fortune 500 companies also fell victim to the breach.

The SolarWinds attack has been a big wakeup call for many organizations worldwide, and the attack was a major new event when it was disclosed to the public in early 2021. Can metaverse applications be impacted by Software supply chain attacks? The answer is a resounding yes, and we need to get prepared and vigilant of this kind of attack.

Metaverse has its dependence on various software and firmware embedded in IoT/AR/VR devices. These software and firmware can easily fall prey to hackers if left unprotected, especially if the metaverse application depends on its supply chain software or service component to function well. If the supply chain software or service is down, the metaverse application will also not function properly. Supply chain risks go beyond supply chain attacks, since in addition to supply chain attacks, which are caused by malicious actors, the supply chain risks include risks due to supply chain service providers' honest mistakes or downtimes.

Three examples of supply chain risks in OpenSea, Mark-DAO, and Lendf.Me are presented here to better explain these risks.

OpenSea Down, Impacting Wallets, and Other NFT Projects

In January 2022, OpenSea, one of the most popular market-places for NFTs, suffered a "database outage." As a result, several services that rely on OpenSea's APIs, including the popular crypto wallet MetaMask, had trouble displaying NFTs.

"We're caching that data so their outage doesn't wipe the wallet," MetaMask co-founder Dan Finlay mentioned in an email. "We store what NFTs the user has in the wallet. Open-Sea's outage means we are currently not auto-detecting new NFTs that are sent to the user's wallet, although users can always tell the wallet about NFTs they have by entering its address manually."

"We use OpenSea to detect new NFTs, this outage would only affect new NFTs minted during the outage. Users can still manually add NFTs to our wallet, and this only affects the auto-detection of NFTs," Finlay said. "Auto-detection (the thing not working) is itself a feature we're going to make opt-in anyway to improve user privacy, so in a way, privacy advocates might prefer the current behavior (i.e., users to manually add NFTs)."

In other words, because OpenSea was down, some NFT owners who just bought their tokens could not see their expensive JPEGs even in their crypto wallet. To be clear, users still "owned" a unique string of characters – or hash – that showed the world they "owned" their expensive JPEGs, and most users were able to view their NFTs just fine due to MetaMask's caching workaround. But some were not able to see new NFTs in MetaMask until OpenSea came back online.

Similarly, in December 2021, an outage at Amazon's Web Services showed that at the end of the day, the web right now is very interdependent and not decentralized at all. When AWS went down, dYdX, a so-called *decentralized exchange* on Ethereum went down as well.

In the OpenSea and AWS cases, we can view Web2.0-based applications acting as supply chain software or services for Web3 applications. The security breaches and attacks or downtime caused by DDOS attack or honest mistakes in Web2.0-based server operations can cause damage to Web3 applications. However, converting Web2.0 to Web3 or developing fully decentralized Web3 applications are currently not possible due to computation power and storage limitations of Web3 infrastructure built on top of blockchain.

Even if you can build a metaverse application with majority of logic implemented in smart contract on a

Web3 infrastructure, you still have supply chain risk issues if one smart contract fails to invoke another smart contract, or a smart contract invoking another smart contract (the so-called "money Lego" or "metaverse Lego" built on smart contract) leading to the unexpected result of losing user's funds. The following two examples indeed show that these two scenarios are possible.

Maker DAO Crash on 2020 "Black Thursday"

The decentralized finance lending platform Maker, like many crypto participants, suffered losses during the price collapse of "Black Thursday" on March 12, 2020. The price of ether (ETH) declined by about 50 percent within 24 hours, triggering opportunistic profiteering as the Maker system became swamped with a huge volume of liquidations.

The reason for the MakerDAO crash of Black Thursday was the failure of Oracle as the price feed. Here is what really happened:

1. **Ethereum network overwhelmed, gas prices increased.** On 12 March 2020, the Ethereum network was overwhelmed by demand as the price rapidly plummeted. The transaction queue grew as network capacity was reached, and gas prices shot up by an order of magnitude.
2. **Price oracles failed.** Due to unusual high gas prices in Ethereum, which can be viewed as supply chain software for MakerDAO, the price oracles including the Maker "Medianizer" failed to update their feeds.
3. **The vault (holding ETH coins) liquidations lagged.** When the Medianizer feed was updated, the reported price instantly decreased by over 20 percent, causing many vaults to be liquidated immediately.
4. **ETH was sold for free through Maker.** Again, due to high gas fees and network congestion, when the ETH collateral in the vaults was auctioned off, many bids did not get

through. This allowed some liquidators to win these auctions with bids of zero DAI (dai is a stablecoin cryptocurrency for MakerDAO) by paying high gas fees, extracting over $8 million worth of ETH, essentially for free.

5. **Vault owners left with millions in losses.** This exploit means that over $4.5 million of dai in the MakerDAO system is unbacked. In addition, users whose vaults were liquidated (and whose ETH was sold to the zero-bid liquidator) lost 100 percent of their collateral, resulting in millions of dollars of losses for the DeFi community.

Following the events of Black Thursday, the Maker community sought to implement protocols that would prevent a situation where keepers were unable to participate in an auction bid. The zero-price and half-price bid exploits only worked because those auctions had only one bidder and were thus able to liquidate ETH with minimum DAI bids.

Unfortunately, after the fallout of MakeDao on Black Thursday, MakerDAO only focused on the ways to limit the zero price, to limit size of bid, and to require the large size bidder to have large capital as collateral. Although these are good measures, these alone will not prevent future issues when the price oracle (as supply chain service) stops to provide accurate price quotes to the system and will again allow arbitrators to leverage flash loans to game the system to get cheap liquidation prices for the collateral assets. A better approach is to have a failover and more redundant solutions for oracle price quotes and improvement in the base blockchain system if Ethereum gets too congested.

Supply Chain Smart Contact Attack on Lendf.Me

On April 19, 2020, Lendf.Me, the lending protocol in the dForce network, was attacked and approximately $25 million in assets were drained from the contract. The hackers utilized a vulnerability in ERC777 tokens as a supply chain smart contract built into the DeFi smart contracts to execute a reentrancy attack. The callback mechanism enabled the hacker to supply and withdraw ERC777 tokens repeatedly before the balance was updated.

What's in the background is that the promise of DeFi system was built on the so-called "money Lego." Each "Lego" in this case is a smart contract file or multiple smart contract files in ".Sol extension." Leveraging these money Legos as base protocol or as supply chain smart contract can help project teams to build very useful and innovative DeFi systems. Unfortunately, the supply chain risk is largely ignored by most DeFi systems. We will continue to see more and more attacks due to supply chain smart contract or money Lego vulnerabilities in newly built DeFi systems or the metaverse systems leveraging DeFi applications. For example, the Wormholes case illustrates the supply chain code risk on the cross chain technology.

Wormhole is a protocol that enables users to move their tokens and NFTs between Solana and Ethereum. As one of the most popular bridges linking the Ethereum and Solana blockchains, Wormhole lost more than $320 million in February 2022. It is the second-biggest exploit ever in the DeFi world, just after the $600 million Poly Network hack, and it is the largest attack to date on Solana, a rival to Ethereum that is increasingly gaining traction in the NFT and DeFi ecosystems.

Crypto holders often do not operate exclusively within one blockchain ecosystem, so developers like Wormhole have built cross-chain bridges to let users send cryptocurrency from one chain to another. The Wormhole protocol used a third party digital signature verification algorithm. The third-party code as a supply chain code was deprecated and could not properly verify digital signatures. The hacker was able to leverage this supply chain code risk to bypass the signature verification requirements and steal over $320 million in value of Ethereum token.

In summary, supply chain software attacks and risks are a real danger to metaverse applications, regardless of Web2 and Web3 technologies used in the applications.

Quantum Computing: Challenges and Opportunities

The metaverse applications will most likely run on blockchain that relies on the security of the cryptographic processes underlying it. Without trusted hash functions and public key signatures, there are no "real" blockchains. Quantum computers, which perform computations deemed impossible with a classical computer, threaten several of the cryptographic primitives used in blockchains. Universal, scalable quantum computers, which are necessary to attack the mathematical problems behind the cryptographic primitives, are not yet available.

However, small-scale quantum computers, with a restricted input size and a restricted number of computations, have already been built by several companies and world governments. Some are even accessible on the internet and can be used to test quantum algorithms. Quantum supremacy, which describes the point in time when quantum computers explicitly outperform classical ones, has either been attained or is on the verge of realization. Below we analyze the key cryptographic primitives in the quantum era.

Random Number Generation

Random number generation is at the core of most cryptographic processes. As classical computers are deterministic, generating good randomness is not easy. There are many instances in which poor randomness led to disaster. This is especially true for blockchain, where random numbers are applied at various levels of the protocol. The problems are more acute with isolated servers, where most of the computations are performed without any human intervention.

Here, quantum technologies can help. Quantum theory is indeterministic by essence. To generate random numbers based on quantum is therefore a safer way to provide good randomness. Quantum random number generators (QRNGs) now exist in very small form factors. (See, for example, the

Quantis QRNG chip from ID Quantique.) Such small QRNGs can be easily integrated into the servers, maintaining the nodes of the blockchain, and even in various users' terminals, such as PCs and smartphones.

Hash Functions

Cryptographic hash functions are heavily used in the blockchain for address generation, proof of work, digital digest, and merkel proof. They transform a text input of any length into a fixed length output. The output is deterministic linked to the input, but it is impossible to recover the input from the output, except by brute force – trying every single input until the correct output is found. The most commonly used hash function, SHA256, has a 256-bit output. A brute force attack on this function would require 2^{256} operations, well beyond the capacity of even the largest supercomputer. A quantum attack with the Grover algorithm would reduce this to 2^{128}, which is still unfeasible for a brute force attack.

The quantum computer will not be able to destroy the immutability of blockchain that is partially protected by hash functions, but it may necessitate a doubling of the hash function size. For example, for the nodes on a PoW (proof-of-work)-based network, they must compute the smallest SHA256 in order to win block rewards. Here the Grover algorithm implemented on a quantum computer will allow a much faster calculation.

Public-Key Signatures

The public-key signatures used in the blockchain are based on Elliptic Curve Cryptography (ECC), which has a very small key size and is easy to implement in the blockchain environment. Unfortunately, it is now known that the current ECC will be destroyed by the Shor algorithm implemented on a quantum computer.

This means any public key published on blockchain may leak the corresponding private key to an adversary equipped with a quantum computer. This is a catastrophe for some blockchains, such as the upcoming new release of POS (proof-of-stake)-based Ethereum 2.0, where publishing the public key is required. It is much less serious for other types of blockchains, such as Bitcoin, where the publicly available address is a hash of the public key, or consortium blockchains, which leverage symmetric-key cryptography.

Transition from Pre-Quantum to Post-Quantum Blockchain

Advances in quantum computing have triggered a growing sense of urgency within the blockchain community to identify post-quantum algorithms that are both effective and practical to deploy. The transition from pre-quantum to post-quantum blockchain is necessary to ensure the security of blockchains in the quantum era.

The extensive use of digital signatures in support of conducting blockchain transactions represents a prime vulnerability. Much attention is therefore being placed on the development and selection of suitable post-quantum digital signature algorithms, which are suitable for blockchain applications and can be phased in over time. Some of the requirements are outlined below:

1. **Some computationally intensive post-quantum cryptosystems may not be suitable for certain hardware currently used for implementing blockchain nodes.** Therefore, post-quantum schemes should provide a tradeoff between security and computational complexity to not restrict the potential hardware that may interact with the blockchain. One possibility is having gradations of key strength based on the hardware available.

2. **Certain post-quantum crypto systems generate large overheads that may impact the performance of a blockchain.** To tackle this issue, future post-quantum developers will have to minimize ciphertext overhead and consider potential compression techniques.

3. **To increase security, some post-quantum schemes may limit the number of messages signed with the same key.** Therefore, it would be necessary to generate new keys continuously, which involves dedicating computational resources and slowing down certain blockchain processes. Therefore, blockchain developers will have to determine how to adjust such key generation mechanisms to optimize both speed and transactions. (In the selection of the right schemes, the NIST Post-Quantum Cryptography Standardization project is widely recognized as the preeminent authority that will drive the selection and adoption of post-quantum algorithms.)

In parallel to security risks, quantum computing can also help metaverse as follows:

- **Security.** Often quantum computing is discussed as a security threat, but if more and more of our interactions are captured in the metaverse, then we will need quantum resistant security for all the transactions and commerce that takes place. Quantum resistant technologies may need to be adopted to ensure that transactions remain safe against algorithms such as Shor's algorithm.

- **Fast computation.** Researchers are developing applications around tasks such as optimization. With massive amounts of computation and simulation required in the metaverse, any advantage that can be leveraged will likely be used to enhance the experience.

- **Randomness.** To create realism, metaverses will need an element of randomness to make sure that hackers and their algorithms do not game the system. One possible way to a high degree of randomness is to utilize quantum randomness, which means that instead of a pseudo random number, a set of qubits can be used to create random bits. Companies such as Quantum Dice are active in this field of QRNG.

In summary, quantum computing is both an opportunity and challenge for the metaverse.

Extended Reality (XR): Novel Security Risks

Extended reality (XR) is a catch-all word for any technologies that improve our senses, whether by delivering extra information about the real world or building completely unreal, simulated worlds for us to explore (see **Figure 8.5**).

- Augmented reality (AR) is an *interactive experience* of a real-world environment where the objects that reside in the real world are enhanced by computer-generated perceptual information, sometimes across multiple sensory modalities, including visual, auditory, haptic, somatosensory,

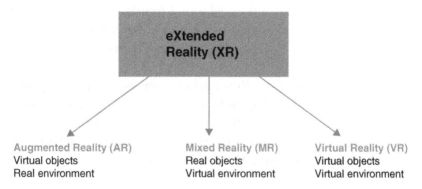

Figure 8.5   Illustration of AR/VA/MR/XR

and olfactory. Snapchat glasses and the game Pokémon Go are two examples of augmented reality experiences.

- Virtual reality (VR) means a completely *immersive experience* that isolates the user from the outside world. Users interact with virtual objects in virtual environments. Users may be transported into various imagined situations using VR devices such as the HTC Vive, Oculus Rift, or Google Cardboard.

- Real-world and virtual digital items interact in a mixed reality (MR) experience, which includes *features of both* AR and VR.

Whilst virtual reality technologies have been available for decades, we are now in a period of rapid growth in XR availability and adoption, which can be attributed to many factors including a reduction in hardware cost, increases in the availability of high-speed, high-quality connectivity, and most recently, shifts in society brought on by the global pandemic (see **Figure 8.6**). Spanning the full breadth of XR, the metaverse will blur the distinction between our online and physical worlds even more.

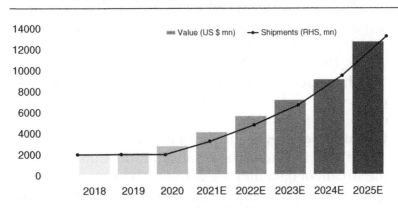

Figure 8.6   XR Smart Hardware Market Grows Rapidly Post-Covid

Source: Credit Suisse, "Global TMT Sector Metaverse: A Guide to the Next Gen Internet," February 2022.

The metaverse is the best of both worlds, where we have the freedom to work and create from a place of our choosing while also capturing the nuances of expression and how we exist in the real world. It's a return to the organic way we interact. But novel security risks are also emerging; for example:

- **"Human joystick" attack.** One potential form of attack, identified by Baggili and other University of New Haven researchers, is what they call the "human joystick" attack. Studied using VR systems, the researchers found that it's possible to "control immersed users and move them to a location in physical space without their knowledge," according to their 2019 paper on the subject. In the event of a malicious attack of this type, the chances of physical harm are heightened.

- **Chaperone attack.** Likewise, a related threat identified by the researchers is the "chaperone attack," which involves modifying the boundaries of a user's virtual environment. This could also be used to physically harm a user. The whole point of these immersive experiences is that they completely take over what you can see and what you can hear. If that is being controlled by someone, then there's absolutely the possibility that they could trick you into falling down an actual set of stairs, walking out of an actual door, or walking into an actual fireplace.

- **Overlay attack.** Additional potential threats identified by the University of New Haven researchers include an "overlay attack" (which displays undesired content onto a user's view) and a "disorientation attack" (for confusing/ disorienting a user).

- **Denial of service.** In this scenario, people who rely on AR displays for their jobs are abruptly disconnected from the stream of data they are getting. This can occur in any application area. However, AR is particularly concerning since many professional employees may utilize the technology to complete jobs in crucial scenarios when a lack of knowledge might have devastating or even fatal repercussions.

- **Untrustworthy information.** Untrustworthy information is also another major concern of AR/VR/XR. Graphics and information are superimposed over the real environment in augmented reality. Gamers, retailers, architects, and professionals will make real-world decisions based on the information offered by AR applications. Hackers might inflict harm if they breach an app and display bogus information and graphical elements on a victim's AR display or glasses. Imagine a doctor using an AR display to check on a patient's vital signs, only to be presented with incorrect information and miss a patient who needs immediate treatment.

- **Theft of XR-related identity.** Criminals may steal AR/VR/XR-related identities associated with real people. Hacking might be a cyber concern for retailers who utilize augmented reality and virtual reality shopping apps. Many consumers' credit card information and mobile payment methods are already stored in their user profiles. Because mobile payment is an easy process, hackers may obtain access to these and secretly deplete accounts.

For AR/VR/MR/XR, the human safety concerns will be the main challenge. As the virtual world meets the physical worlds, XR is the connector and without security and safety design, the cost could be physical damage or even life threatening. Imagine a possible metaverse experience in the future when a person rides a self-driving car and plays an NFT game inside the car and the hacker gets control of the car or the XR headset to crash the car directly or indirectly into another car, causing a life-threatening car accident. This is possible if we do not have security and safety design in XR-powered metaverse applications.

To conclude this chapter, the Web3 metaverse could bring tougher security risks than the Web2.0. Today's threats may still exist, but there is also the potential for newer, novel threats.

For example, fraud and phishing attacks could come from an avatar designed to impersonate a coworker instead of a misleading domain name or email address. Also, the tech stack of the Metaverse, which includes eye-tracking and biometrics, means our senses are also being monitored and analyzed with increasing accuracy.

Therefore, we need to learn from the mistakes of Web2.0 and be proactive about Web3 security. This is critical at a time when numerous sectors, both at the enterprise and consumer levels, have begun adopting the metaverse for ubiquitous, holistic, and fully immersive experiences. With the help of blockchain technology, we have one chance at the start of this era to establish specific, core security principles that foster trust and peace of mind for metaverse experiences.

PART

Three-Way War among Open Metaverse, Big Tech Walled Gardens, and Sovereign States

Just like in the context of digital currency, where a three-way competition among the cryptocurrencies, Big Tech tokens, and CBDCs (central bank digital currency) intensifies, for Web3 infrastructure, the open Metaverse must compete with both Big Tech corporations and government-backed blockchain networks.

Public Crypto, Government CBDC, and Big Tech Coin

- Three-Way Currency War in Metaverse(s)
- China's eCNY Push at 2022 Winter Olympics
- Crackdown on the World's Largest Crypto Market
- Digital Rupee, Digital Ruble, and Britcoin
- US Bellwether: CBDC R&D and Crypto Regulation
- US-China Consensus: Stablecoins in the Regulatory Spotlight
- Big Tech Coin: The Rise (and Fall) of Libra

Three-Way Currency War in Metaverse(s)

In January 2022, the Polynesian island Kingdom of Tonga was hit with a tsunami in the aftermath of a giant volcanic shockwave. According to a *Cointelegraph* report, the tsunami hit the citizens of Tongatapu, the main island of Tonga, and the news inspired cryptocurrency holders to want to donate bitcoin for disaster relief for those affected by the tsunami. In true bitcoin fashion, it was on Twitter that user Onair Blair urged former Tongan lawmaker Lord Fusitu'a to set up a bitcoin wallet address for crypto holders around the world to donate for tsunami relief purposes. Lord Fusitu'a obliged.

In the same *Cointelegraph* report, Lord Fusitu'a discussed Tonga's plan to use geothermal energy from Tongan volcanoes (21 in total) to fuel bitcoin mining. The mining effort would benefit Tongan finances, according to Lord Fusitu'a: "Each [Tongan] volcano produces 95,000 megawatts at all times leaving much to spare, and a single volcano can generate $2,000 worth of bitcoin daily, which will be given to Tongan families." Furthermore, Lord Fusitu'a, a proponent of Bitcoin and a Tongan politician, stated to *Cointelegraph* that he wished to make bitcoin legal tender in Tonga, just as El Salvador has.

The case of Tonga illustrates the widening of public acceptance of cryptocurrency – not only in the virtual cyberspace, but also in the physical world. Unlike our current Web2.0 world, which centralizes communication and commerce within proprietary platforms like Google and Amazon, Web3 is expected by its supporters to be open and decentralized. Meanwhile, platform decisions would be executed through transparent smart contracts (autonomous software programs) among the Web3 enthusiast. How would you pay for everything in Web3? With cryptocurrencies, of course. Communities of Web3 users are expected to maintain and operate applications and services in exchange for cryptocurrencies like bitcoin, gaming tokens, or NFTs.

That's why developers are building all kinds of financial applications on the blockchain, potentially enabling crypto-based payments between virtual and physical worlds (see **Figure 9.1**). For example, early 2022 Solana Labs announced the launch of the Solana Pay, which may enable merchants to accept cryptocurrencies payments directly from consumers. It's important to note that this is bigger than enabling consumers to "pay with crypto." Rather, this is about a vision where all currencies – including US dollars – are on-chain and used for a wide range of transactions.

The background is that merchants have been able to accept crypto currencies for years (e.g., Tesla accepted bitcoin

Three-Layer Architecture of the Metaverse

Global Payment Rail is the Link for the Virtual World and Physical World

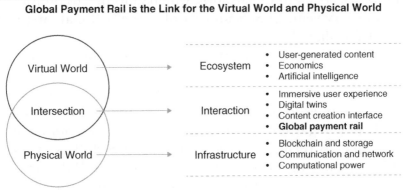

Figure 9.1   Crypto-Based Payments link Virtual and Physical Worlds
Source: Duan et al, Crypto.com Research; Data as of 18 Nov 2021

for purchase of its cars), but acceptance usually means swapping out one intermediary for another. For Solana Pay, it is about a vision where all currencies – including US dollars – are on-chain and used for a wide range of transactions. The core premise behind Solana Pay is that the payment and underlying technology goes from being a necessary service utility to a true peer-to-peer communication channel between the merchants and consumers.

However, there are headwinds from the existing financial establishment. For example, the Central American El Salvador made headlines in September 2021 after it became the first country to adopt bitcoin as legal tender, but its implementation process has not been smooth.

The move to adopt Bitcoin was heavily championed by the country's president, Nayib Bukele. According to news reports, Bukele stated that El Salvador legalized Bitcoin for a couple of reasons: increasing investment into El Salvador, improving financial access for 70 percent of Salvadorans who do not have access to "traditional financial services," and giving expatriate Salvadorans an easier way to transfer remittances back to El Salvador, which amounts to billions of dollars every year. El Salvador's only other official currency is the US dollar.

It can be argued that this move is both innovative and controversial. Besides the tech implementation issues as the Bitcoin operations went live, many organizations around the world have questioned the soundness of this act. Because bitcoin is now legal tender, merchants across the country may be obliged to receive bitcoin instead of US dollars, a more stable currency in terms of price volatility. The IMF, for one, has criticized El Salvador for its bitcoin legalization, on the grounds that the risks of using the volatile cryptocurrency (for reference, bitcoin lost half its value from November 2021 to January 2022) will severely endanger the financial stability of the country.

The IMF has stressed that if Bukele does not move to delegalize bitcoin, El Salvador will find it difficult to procure loans from the IMF given its risky financial status. Furthermore, the IMF issued a January 2022 press release where it urged El Salvador's authorities to narrow the scope of the Bitcoin law by removing bitcoin's legal tender status.

The World Bank has reacted in a similarly critical fashion. When El Salvador approached the financial institution in 2021 for assistance in implementation of bitcoin as legal currency, the World Bank rejected its request. A June 2021 BBC report quoted the World Bank stating its "concerns over transparency and environmental impact of Bitcoin mining." The question the World Bank has raised dovetails into the other risk of bitcoin legalization for El Salvador: crypto mining.

El Salvador, also known as "the Land of Volcanoes," may get another use for its volcanoes because of its bitcoin legalization initiative. In November 2021, President Bukele announced that he plans to build "a Bitcoin city at the base of a volcano (the Conchagua), using bitcoins to fund the project," according to a BBC report. Bukele noted that the city will take advantage of the Conchagua volcano's geothermal energy to power Bitcoin mining. Whether El Salvador is able to pull off the crypto mining effort without severely damaging the environment around the Conchagua is difficult to predict, as international observers such as the World Bank anxiously await.

Furthermore, the government of major economies are not sitting idle to watch cryptocurrencies rising. Globally the central bank, which is usually a conservative fortress, is also breaking new ground. As more economic activity moves online (especially after the Covid-19 disruption) and physical cash is at a disadvantage, many are on the road to introducing their own digital currencies – the central bank digital currency (CDBC). A 2021 BIS survey of central banks found that 86 percent are actively researching the potential for CBDCs, 60 percent were experimenting with the technology and 14 percent were deploying pilot projects.

In summary, it remains too early to call crypto the "default currency" in the future Metaverse (see **Figure 9.2**). The crypto ecosystem must fight a currency war on two fronts. On one hand, the war against the financial establishment of governments, including the national CDBCs, which competes

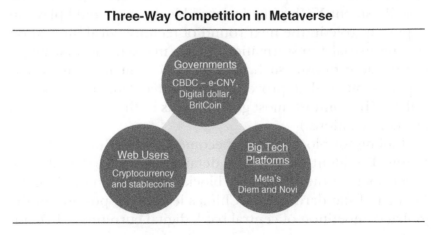

Figure 9.2   Three-Way Currency War in Metaverse

with cryptocurrencies for transactions in the Metaverse, and crypto regulations that limits the usage and trading of crypto assets.

On the other hand, Big Tech companies such as Meta are trying to provide unprecedented expansion of financial products on a single platform, facilitated by Meta's own coin (Diem) and digital wallet (Novi). This chapter will first discuss CBDC development and crypto regulations from the major economies such as China, US, India, and Russia, followed by the analysis of the coins from Big Tech companies like Meta.

China's e-CNY Push at 2022 Winter Olympics

Today, China and the United States are competing and growing their technological capabilities in a wide array of sectors. The frontier of America and China's technological war is around who will dominate the blockchain and cryptocurrency industry.

Although China has cracked down on cryptocurrencies (details in the following section), shutting down all domestic crypto exchanges and banning all ICOs (initial coin offerings), the government recognizes blockchain technology itself as a revolutionary development. In an October 2019 speech, Chinese President Xi Jinping declared blockchain would play "an important role in the next round of technological innovation and industrial transformation." That marked the first major world leader to issue such a strong endorsement of the widely hyped – but still unproven – distributed ledger technology (DLT). (By contrast, most governments in the West have been far more cautious.)

Calling for blockchain to become a focus of national innovation, President Xi's speech detailed the ways the Chinese government would support blockchain research, development, and standardization. China's leadership position in the global competition of central bank digital currency (CBDC) is

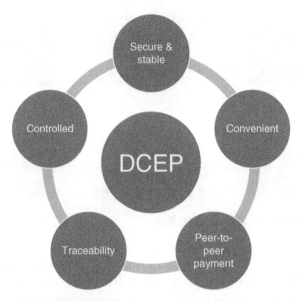

Figure 9.3   How Is DCEP Designed?
Source: Media Reports, April 2020

the prime example. Unlike bitcoin and other cryptocurrencies built on the excitement regarding "decentralization," China's CDBC, which is called e-CNY by the People's Bank of China (PBOC), China's central bank (CNY is the English synonym of the Chinese currency, the yuan), is run on a centralized database; nevertheless, e-CNY is built with blockchain and cryptography, and it has incorporated blockchain's key concepts such as peer-to-peer payment, traceability, and tamper-proof-ness (see **Figure 9.3**).

The e-CNY's timeline began years after the development of bitcoin and other cryptocurrencies (see **Figure 9.4**). The first publicized PBOC effort on digital currency occurred four years after the first bitcoin transaction in May 2010, in the form of the establishment of an in-house digital currency research group. Starting 2017, PBOC accelerated its efforts and first established the Digital Currency Research Institute in January

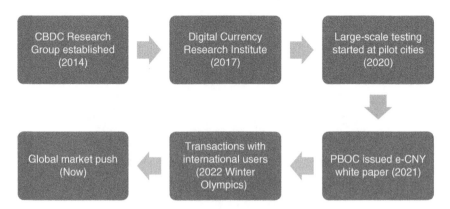

Figure 9.4   Timeline of China's Digital RMB

2017, then followed with December 2017's call for Chinese commercial banks and payment institutions to collaborate in efforts in digital currency. Further acceleration of e-CNY testing was broadly viewed as a reaction to Facebook's announcement in 2019 that it intended to launch *Libra*, Facebook's planned blockchain-based digital currency (see detailed discussion later in this chapter).

The large-scale testing occurred in 2020 at numerous major cities amid the Covid-19 pandemic. In those pilot zones, e-CNY has been formally adopted into the cities' monetary systems, with some government employees receiving their salaries in the digital currency from May 2020. People can create an e-CNY wallet in their commercial banks' mobile app and use the national digital currency for things like transportation, education, healthcare, and other consumer goods and services. Starbucks, McDonald's, and Subway chains in China, for example, were named on the central bank list of firms to test the digital currency.

The rapid development of the e-CNY is only an inevitable result of China's commitment to digital transformation. In the last several years, mobile payment has grown rapidly to become the dominant form of payment in China. (Reference to Winston Ma's 2016 book *China's Mobile Economy: Opportunities in the Largest and Fastest Information Consumption Boom*.) Since 2020, China has been steadily expanding its digital yuan pilot programs (while also cracking down on cryptocurrencies; see detailed discussion in the following section), given the country's rapid development of internet industries such as e-commerce and social network platforms that provide a myriad of application scenarios.

In July 2021, the PBOC issued a white paper detailing the current workings of the digital yuan, also referred to as the e-CNY, which is the first comprehensive disclosure of its plans. The release of the white paper probably marked the near end of the testing phase for the digital currency's "2C" retail payment. The digital yuan wallet supports several functions, including scan to pay, top-ups, and money transfers. According to the white paper, as of June 2021, participants have spent 34.5 billion digital yuan ($5.3 billion) in trials. Uses included paying utility, dining, transportation, shopping, and government services. In January 2022, the eCNY wallet was the most downloaded app in Apple and Xiaomi App stores within just a week of formal launching.

The new digit yuan would allow users to spend it even without an internet connection, and it will bring convenience to foreigners, too. "Foreign residents temporarily traveling in China can open an e-CNY wallet to meet daily payment needs without opening a domestic bank account," said the white paper. That means even foreigners traveling in China can have access to the digital yuan without a domestic bank account. This is a particular benefit given the difficulties that foreigners have had using mobile payment apps like WeChat Pay (of Tencent) and Alipay (of Alibaba), because those apps must be linked to banking accounts.

In February 2022, the Beijing Winter Olympics became a major milestone for China's digital currency because it was the first test for the digital yuan with international users. It was also like a stress test for the digital currency infrastructure because the system handled the exchange of foreign currencies from numerous countries. For example, global athletes and visitors could put their own money, from different countries, into ATM machines and then get a debit card in Chinese digital yuan, converted from the foreign money. With the debit card, they could go to restaurants and go shopping nearby, without needing a banking account in China.

Of course, the profound impact of e-CNY is likely to be more than China's retail markets. Most likely, e-CNY would make China the first major economy to adopt a native digital currency. Many believe the e-CNY will bolster Chinese currency's global status and eventually challenge the US dollar's preeminent position as the world's reserve currency. For example, the e-yuan could bypass Western-operated cross-border payment networks, such as SWIFT, which the United States has used to enforce sanctions.

But it's likely a long march for the e-CNY. "Though technically ready for cross-border use, e-CNY is still designed mainly for domestic retail payments at present," the e-CNY White Paper reads. For the e-CNY, its real test only starts after its official launch in 2022, and in the global context e-CNY's competitors are rising, as its progress has stimulated western countries and advanced economies to follow suit and modernize their financial systems. Later in this chapter we will see many countries starting to develop their own CBDCs, such as digital dollar, digital rupee, and digital ruble.

Crackdown on the World's Largest Crypto Market

Before China's State Council's Financial Stability Committee vowed to crack down on the cryptocurrency's mining and trading activities in May 2021, few people – even among global financial professionals – realized that China accounts for more than 70 percent of the world's bitcoin and other cryptocurrencies' supply. Because most global cryptocurrencies were mined and traded in China, Chinese regulations in this new industry have profound global implications.

The 2021 crackdown is not the first time China has strengthened regulation of cryptocurrencies. China issued similar bans first in 2013, and then in 2017, when China accounted for 90 percent of global bitcoin trading. The 2017 rule issued by China's central bank, the People's Bank of China (PBOC), and other ministries, essentially shut down local cryptocurrency exchanges, forcing major exchanges including Binance and Huobi to relocate overseas.

Nevertheless, onshore Chinese investors could still trade cryptocurrencies on platforms owned by overseas exchanges. As the price of bitcoin jumped multiple times since late 2020, Chinese trading activities also heated up.

As such, the May 2021 crackdown was viewed by the cryptocurrency market as just another rule announcement without serious enforcement. For example, Hong Kong's Bitcoin Association said in a tweet in response to China's reiterated ban: "For those new to bitcoin, it is customary for the People's Bank of China to ban bitcoin at least once in a bull cycle."

But this time is different. Coming from the State Council's Financial Stability Committee, the highest level financial regulator of China led by vice premier Liu He, the new cryptocurrency crackdown is a significant upgrade of existing regulations. Furthermore, it is the first time the State Council has explicitly targeted cryptocurrency mining activities, which indicates a determination to crack down on cryptocurrency trading from its origin, as China is the largest cryptocurrency mining field in the world.

The Chinese government has suggested that investor protection, carbon neutrality, and financial stability are the three key factors for the new regulations. The regulatory development of China, the largest cryptocurrency mining field and trading market in the world, will be an important reference case for other countries that start developing regulations for the cryptocurrency mining and trading activities.

Investor Protection

Investor protection – cutting off the cash flow channel between uneducated investors and offshore exchanges – is a motivation for new regulations. For the Chinese regulators, bitcoin and other cryptocurrencies are not investment tools; rather, they are speculative instruments with high volatility. China has a clear record of cracking down on all kinds of products for fear that bubbles will eventually burst and lead to riots of disgruntled retail investors – whether it is in beans, garlic, tea, or the more recent, peer-to-peer loans.

Since the State Council's decision in May, three Chinese financial associations – the National Internet Finance Association of China, the China Banking Association, and the Payment and Clearing Association – have issued a new rule to ban financial institutions from cryptocurrency-related businesses. The rule is designed to make it more difficult for individuals to buy cryptocurrencies using various payment channels. The associations have reminded investors that virtual currencies "are not supported by real value."

To ensure that all the rules will be seriously enforced, the PBOC summoned representatives of multiple institutions, including state-owned commercial banks and Alipay, and told them to "strictly implement" recent notices and guidelines from authorities on curbing risks tied to cryptocurrency transactions. As China-focused exchanges that are registered overseas allow Chinese individuals to open accounts online, and cryptocurrency transactions by Chinese individuals can be made through banks, or online payment channels such as Alipay or WeChat, the financial firms were also instructed to go through their systems to investigate and identify customers with accounts in virtual currency exchanges, in which case, the institutions have to cut off the accounts' ability to send or receive money for transactions.

Carbon Neutrality

Another motivation for new cryptocurrency regulations is China's goal toward carbon neutrality. China's new environmental policy is a key factor in the mining crackdown and was not part of previous cryptocurrency regulations. President Xi Jinping, in a November 2020 speech to the UN General Assembly – months before the cryptocurrency crackdown – pledged to have the nation's carbon emissions peak before 2030 and realize carbon neutrality by 2060.

The carbon neutrality policy cuts back coal power, which has been a major energy source for the country. According to London-based climate data provider TransitionZero, China needs to halve its carbon dioxide emissions from coal-based power plants by 2030 to achieve the policy. To meet climate targets, cryptocurrency mining is one of the focus areas, as it is one of the many high-energy-consumption industries in China. After the central government initiated the cryptocurrency crackdown campaign in May, major coal-based power producers such as Inner Mongolia and Xinjiang, which were previously the top two cryptocurrency mining hubs in China, have been among the first regions that quickly developed local rules to clean up mining businesses.

Furthermore, China's carbon neutrality policy created an energy shortage within the country due to its drastic reduction in coal-fired power, which means that even mining with renewable energy, like hydropower, is subject to new regulations. Sichuan and other provinces also had to shut down all mining businesses in June, whether they were powered by coal or hydro.

Financial Stability

A third motivating factor for cryptocurrency regulation is to maintain financial stability as well as to push forward China's central bank digital currency. As mentioned, in July 2021, the PBOC issued a white paper on its development of China's digital currency, the e-CNY. In its white paper, the PBOC cited the rapid growth in cryptocurrencies as a driver for research and development of the e-CNY and said, "Cryptocurrencies are mostly speculative instruments, and therefore pose potential risks to financial security and social stability."

This white paper description is the first time that the PBOC, in an official document, linked its sovereign digital currency issuance with cryptocurrencies' potential challenges to the international monetary system. According to the PBOC, "cryptocurrencies' lack of intrinsic value, acute price fluctuations, low trading efficiencies and huge energy consumption make them unfit for use in daily economic activities." (Interestingly, that view is not that different from that of US Federal Reserve Chairman Jerome Powell, who at a panel hosted by the Bank for International Settlements (BIS) considered cryptocurrencies "more a speculative asset that's essentially a substitute for gold rather than for the dollar.")

Just like Chinese digital currency e-CNY, China's new crypto regulations have already made a significant impact in the global cryptocurrency markets. First, China's mining crackdown has forced a seismic shift in bitcoin mining patterns. By July 2021, bitcoin's network hash rate, a measure of its computational horsepower, had dropped about 50 percent since its peak level in May 2021. In the end, most of China's bitcoin mining capacity is set to be shut down, with some of the capacity relocating to overseas markets, such as the US and Kazakhstan. Second, from a cryptocurrency trading perspective, China's tightened regulations and enforcement have contributed to bitcoin's price dropping about 50 percent from its all-time high price within a few months.

Finally, and probably most importantly, China's new regulatory framework may influence many countries' cryptocurrency-related regulations going forward. Since China's crackdown in May, countries across Asia, Europe, and the Americas have started their regulatory actions on cryptocurrency transactions and related exchanges. For many countries, China has become a reference case when they consider their own CBDCs and crypto regulations.

Digital Rupee, Digital Ruble, and Britcoin

China's digital currency and crypto regulation framework has influenced many countries' lawmaking in the same fields. For example, India's legislature once considered a complete ban of crypto mining and trading like China did. Finally, India made formal recognition of the role cryptocurrencies play in the country via a backhanded move: a tax imposition. In February 2022, India's Finance Minister Nirmala Sitharaman announced in her union budget for 2022–23 that there will be both a 30 percent tax on any income from the transfer of virtual digital assets and also a 1 percent tax deducted at source on payments made for the transfer of digital assets. No deductions and exemptions from this tax will be allowed.

This 30 percent tax is significant, but the Indian regulation would imply that the government plans to at least include cryptocurrency into the financial ecosystem under the rule of law, instead of targeting it someday as an illegal instrument. More in the fray for Indian cryptocurrency are the government's plans for an Indian CBDC. Indian finance minister also announced in February 2022 that the Reserve Bank of India would formally introduce the bank's digital currency – digital rupee – in the next financial year. What the digital rupee might mean for the future of cryptocurrency in India, along with the existence of paper cash, is still to be determined. It's likely that when India accelerates the development of its CBDC like China, it will also further tighten the regulation of crypto assets.

Russia is another pertinent example. In Jan 2022, Russia's central bank issued a major report to propose banning cryptocurrency use and mining on Russian territory, which came as a shock to the global crypto world, as Russia is the world's third-largest crypto mining country (behind the United States and Kazakhstan, after China ceased to be the world's largest mining nation after its crypto crackdown in 2021) and even legalized cryptocurrencies in 2020. However, Russia has had an extensive past record of arguments against cryptocurrencies, including environmental downsides to mining bitcoin, as well as potential and real risks to the financial stability of Russia.

Russia Central Bank's proposal to ban cryptocurrency mining, trading, and storage has not just domestic but also international implications. Russia's Central Bank estimated that Russian citizens' transactions using decentralized cryptocurrencies amount to $5 billion per year. If the Russia-based mining activities and transaction volume is completely phased out, that would create a sizable dent in the global markets. Furthermore, the Bank is also planning to issue its own CBDC (digital ruble), a move to maintain control over its own financial system.

However, there might still be hope for crypto in Russia. In the last week of January 2022 according to a Coindesk report, Russian President Vladimir Putin called for the central bank and the Russian finance ministry to come to a consensus following the former's call for a complete ban on cryptocurrencies. Putin noted that while the central bank is correct to be concerned with the risks of cryptocurrencies for Russian citizens, policy decisions on crypto should be "offset against certain competitive advantages that Russia holds when it comes to mining, due to the country's surplus of electricity and 'well-trained personnel.'"

The dust has not yet settled on whether crypto has seen its last days in Russia (the Russia–Ukraine war started in February 2022 disrupted that debate), and perhaps there still will be a possibility that crypto can coexist with the Russian economy, after more dialogue occurs between government agencies. Furthermore, just like China (and India), Russia's CBDC (digital ruble) development may sooner or later influence its cryptocurrency policy consideration from the "sovereign digital currency" perspective, in addition to financial stability and environment considerations.

Given cryptocurrency's disruptive nature and ability to circumvent official institutions, most nations share similar concerns over tax evasion, money laundering, terrorism financing, and its impact on monetary policy. However, not surprisingly, the Eastern and Western nations' approaches and strategies with CBDC and crypto are markedly different, reflecting dissimilar political and financial systems and views on privacy and central control. For example, in the UK, the Bank of England and Treasury planned for a CBDC named "Britcoin." In an April 2021 statement to a finance industry conference, UK Finance Minister Rishi Sunak announced the creation of a Bank of England-Treasury taskforce set on "coordinating exploratory work on a potential central bank digital currency."

However, in January 2022, the UK's House of Lords voted "no" to Britcoin's launch, citing a variety of concerns from "far-reaching consequences for households, businesses, and the monetary system for decades to come." In a City AM News report, the House of Lords detailed that a Britcoin rollout would include numerous risks, such as the Bank of England's surveillance of British people's spending choices, potential financial instability when economic downturn prompts people to fast convert to CBDCs, and potentially an over-focus of power for the Bank of England, centralizing an attack point for hostile agents who wish to attack the state. Still, such a vote won't stop the Bank of England looking into the "Britcoin," and the digital version of the UK pound may be launched in the near future.

US Bellwether: CBDC R&D and Crypto Regulation

Of course, the policy and regulatory direction in the United States will be the bellwether for the Western nations. For both

US CBDC development and crypto regulation, year 2022 is a major milestone. For the US CBDC ("digital dollar"), in January 2022, the Federal Reserve released a long-awaited report regarding the potential launch of the US CBDC. The paper discussed the advantages and disadvantages of a US CBDC, with the first step to launch to begin holding extensive dialogue with relevant stakeholders on the issue of the digital Dollar. However, the paper does not commit the Federal Reserve to any policy or design choices.

Overall, the Federal Reserve takes on an extremely cautious approach to the digital dollar like the UK's House of Lords. Pros of the US CBDC, as noted by the central bank's researchers, include: "providing a safe, digital payment option for households and businesses as the payments system continues to evolve, and may result in faster payment options between countries." Meanwhile, the researchers also denoted that there is a significant host of risks that come along with a CBDC's deployment, including the CBDC's effect on: "... the financial sector, the cost and availability of credit, the safety and stability of the financial system and the efficacy of monetary policy." The paper contributes to the public debate but makes clear that the Federal Reserve is unlikely to launch a digital dollar in the near- or medium-term future.

For crypto regulations, by 2022 consensus had emerged among US policymakers, regulators, and the industry on the need for sound regulation of digital assets that supports innovation and inclusion. In the past years, the fragmented US financial regulatory environment and the need for coordination across many federal and state regulators, together with the lack of clear regulatory authority over some types of digital asset activities, has resulted in ambiguity in the regulation of these activities.

Nevertheless, thoughtful legislation remains unlikely to be taken up in the near term of 2022, given Congress's other priorities, crypto industry' increasingly powerful lobbying voice, and the challenges in developing a path forward. Among the relevant regulators in the fragmented US financial regulatory environment – the Federal Reserve, OCC, and FDIC (three banking regulators), as well as CFTC (commodities) and SEC (securities), the SEC has been the most active regulator of the crypto trading market. So far, much of the early action on digital assets by the SEC under the current Chair Gary Gensler was focused on enforcement rather than regulation.

For example, DeFi is a priority for SEC regulation in 2022 because of its spectacular growth and the particular regulatory challenges it represents. The total value locked (TVL) in DeFi protocols increased to roughly $300 billion at the end of 2021, up from roughly $20 billion at the end of 2020, according to defillama.com. In the absence of DeFi-specific regulatory guidance or legislation, DeFi regulation in 2022 may continue to develop predominantly through SEC enforcement. For example, in August 2021, the SEC announced its first enforcement action that it billed as DeFi-related, finding that two individuals had unlawfully offered and sold unregistered securities through the use of smart contracts to sell digital tokens.

Meanwhile, areas of focus for SEC's DeFi actions will likely include unregistered facilities or marketplaces. The SEC's 2021 investigation on Uniswap (decentralized exchange) was a milestone moment, and SEC Chairman Gensler has urged cryptocurrency exchanges to voluntarily register with the Commission. Also, the SEC could expand the definition of an exchange as a backdoor way to regulate cryptocurrency exchanges.

Furthermore, with the market for nonfungible tokens (NFTs) exploding to more than $40 billion, the SEC may also look into the space and find potential intersections with securities laws (NFTs) and exchange regulations (NFT marketplaces). The decentralized exchanges are at the center of the creation of crypto products in the same way stablecoins are at the center of the transaction of crypto products. With the SEC looking at the exchanges and US Treasury looking at stable-coins, the US crypto industry may soon embrace a completely new regulatory framework.

The US government's CBDC and regulatory actions accelerated after President Joe Biden's March 9, 2022, executive order asserting that technological advances and the rapid growth of crypto markets "necessitate an evaluation and alignment of the US government's approach to digital assets." The order mandates multiple reports and studies, and it tasks an alphabet soup of government agencies with responsibility for the effort. (Probably to the surprise of financial regulators, the campaign will be led directly out of the White House.)

According to the order, the Biden administration "places the highest urgency on research and development efforts into the potential design and deployment" of a US central bank digital currency (CBDC). This is a clear statement of support from the administration and a change in tone from the recent Federal Reserve commentary, when Fed Chair Powell released the discussion paper on the digital dollar only two months ago. (Maybe a coincidence – when China starts the international testing of its CBDC (eCNY) at the Winter Olympics in February 2022, Boston Fed and MIT released a report on the open-source code that they have developed and could be used as the groundwork for a CBDC.)

From the regulatory aspect, a key takeaway from the order is that the Administration is not handing over responsibility for national crypto policy to the banking, securities, and other financial services regulators. By giving a seat at the table to agencies such as the State Department, the Domestic Policy Council, the Council of Economic Advisers, the Office of Science and Technology Policy, the Office of Information and Regulatory Affairs, and the National Science Foundation, the order signals a point of view that the potential impact of crypto technology on the US economy, national security, and global leadership indicates that legacy regulatory structures need to be revisited within a broader frame of reference. (Indeed, the assistant to the president for National Security Affairs and the assistant to the president for Economic Policy are responsible for coordinating the work for the executive agencies required by the order.)

What may come out first from all potential US regulatory actions is a new regulation for stablecoins – cryptocurrencies tied to fiat currencies like the US dollar. In the fall of 2021, the Biden administration tasked Congress with coming up with a framework for regulating stablecoins. The administration recommended that only banks be allowed to issue stablecoins while directing regulatory agencies to use their existing authorities to regulate stablecoins as best they can right now. Such a stablecoin regulation approach is likely to be the consensus of different nations, including the US and China.

US–China Consensus: Stablecoins in the Regulatory Spotlight

The US and China don't agree on much these days. But there's one issue on which both superpowers see eye to eye: the regulation of stablecoins, a special type of crypto assets that pegs its value to conventional money.

On July 16, 2021, US Treasury Secretary Janet Yellen called on the President's Working Group (PWG) to develop a regulatory framework for cryptocurrencies. Specifically, Yellen pushed financial regulators to draft stablecoin rules, due to its "potential risks to end-users, the financial system, and national security." The PWG meeting was promptly held on July 19, the following Monday, and it announced the plan to issue recommendations about stablecoin regulations within the next few months. "The secretary underscored the need to act quickly to ensure there is an appropriate US regulatory framework in place," the Treasury reported.

It may be a coincidence, but on the same July 16, the People's Bank of China (PBOC, China's central bank) issued a white paper on its development of China's digital currency (e-CNY), where the PBOC cited the rapid growth in cryptocurrencies, especially global stablecoins, as a driver for its research and development of e-CNY.

"Some commercial institutions even plan to launch global stablecoins, which will bring risks and challenges to the international monetary system, payment and clearing system, monetary policies, cross-border capital flow management, etc.," said PBOC in the white paper. This is the first time that China's Central Bank, in an official document, links its sovereign digital currency issuance with stablecoins' potential risks and challenges to the international monetary system.

Why are stablecoins so important? For a comparison, bitcoin is exciting: its price swoops and dives. Such volatility has made bitcoin well known to the public. But the stablecoins are the opposite, which are crypto tokens pegged or linked to the value of fiat currencies. Because they are boring, they are equally useful: these stablecoins are embedded in crypto trading and lending platforms. How do people trade paper dollar for crypto assets (or crypto-to-crypto)? Usually, they use stablecoins as the medium.

In July 2021, nearly three-quarters of trading on all crypto trading platforms occurred between a stablecoin and some other token. Less well known to the public, the existing stablecoin market is worth more than $110 billion, including four large stablecoins – some of which have been around for seven years.

Stablecoins can be a bridge between two worlds that weren't designed to mix – crypto assets and traditional finance. They make it easier to move funds in traditional currency onto crypto exchanges. Many exchanges don't have the relationships with banks needed to offer regular currency deposits or withdrawals, but they can and do accept stablecoins such as Tether (also known as USDT, the most popular stablecoin).

As such, Tether is especially useful in the China crypto market, because it is the critical link for onshore Chinese investors, whose funding in Chinese RMB (CNY) is separate by Chinese regulators from the offshore USD market, to trade cryptocurrencies on platforms owned by overseas exchanges.

Because the stablecoins are at the center of the global crypto ecosystem, the corresponding regulation is equally important. US Treasury Department actions and China PBOC white papers are echoed by US Securities and Exchange Commission (SEC) Chairman Gary Gensler in his speech at the Apsen Security Forum in August 2021. Chairman Gensler said cryptocurrency markets were "rife with fraud, scams and abuse" and called on Congress to give his agency new regulatory powers. He also singled out stablecoins and explained the necessity of regulation from financial security and securities law perspectives (which also explains why the crypto talk happened at the Apsen Security Forum).

First, from the financial security (and national security) perspective, "the use of stablecoins on these platforms may facilitate those seeking to sidestep a host of public policy goals connected to our traditional banking and financial system: anti-money laundering, tax compliance, sanctions, and the like," said Gensler. The worry here is that the growing size of stablecoins has created a situation where huge amounts of US dollar-equivalent coins are being exchanged without touching the US financial system.

Second, from the securities regulation perspective, "these stablecoins also may be securities and investment companies." To the extent they are, the SEC "will apply the full investor protections of the Investment Company Act and the other federal securities laws to these products."

A third consideration, which has more to do with the US Treasury and Federal Reserve, is the stability (or the lack) of the balance sheet of those stablecoins' issuers. Lawmakers and regulators have expressed alarm that retail investors are not fully protected, should one of the stablecoin firms not have the backing they purport to have.

A useful comparison is with money-market funds, which were created in the 1970s to circumvent rules limiting the interest banks could pay depositors. After promising to maintain the value of their shares at a dollar, money-market funds blew up in 2008 in the global financial crisis. American taxpayers stepped in to forestall a fire sale of their assets and a crash in the market for commercial paper, on which the real economy depends. A collapse of stablecoins could look similar according to banking industry experts.

The reality is that few stablecoins say much about their balance sheets. "They are like money funds, they're like bank deposits, and they're growing incredibly fast but without appropriate regulation," US Federal Reserve Chairman Powell said in testimony before Congress. Tether recently made available an assurance opinion by an independent accountant confirming that all Tether tokens are fully backed by Tether's reserves but disclosures of the breakdown of its assets fall far below the standards expected of a bank. And compared to a money market fund that typically invests all assets in "cash and cash equivalent," of the assets backing the Tether tokens in March only about 5 percent were cash or Treasury bills, according to its public disclosures, and most of the assets were riskier–about half of them commercial paper.

Meanwhile, Circle, the backer of USD Coin (USDC), the second-biggest stablecoin, intends to "become the most public and transparent operator of full-reserve stablecoins in the market today," according to Circle founder Jeremy Allaire's twitter in summer 2021. But USDC's public disclosure released in July 2021 was still a simple attestation as to USDC's reserve assets on a single day way back in May. This is merely a monthly verification of the issuer's bank balance. Minutes after the attestation, the stablecoin issuer could simply transfer funds elsewhere.

In summary, it may be stablecoins' turn in the regulatory spotlight. China's crypto mining and trading crackdown started in May was like the tipping point that urged regulators across the globe to accelerate crypto-related lawmaking in their own markets. PBOC's digital currency white paper in July is China's official start of stablecoins regulation, and it may stimulate similar regulations elsewhere in the world.

In the US, at the end of July 2021 US Congressman Don Beyer started pushing the "overdue process" of updating crypto regulations, including a provision to potentially provide the US Treasury Secretary the authority to permit or prohibit US dollar and other fiat-based stablecoins. At the same time, the new bill aims to provide the Federal Reserve with explicit authority to issue a digital version of the US dollar, which may further stimulate the US to develop a "comprehensive legal framework" for crypto assets.

Given the common focus on stablecoins by the US Treasury department, Federal Reserve, the SEC, and the legislature, the regulation of stablecoins may emerge soon in the US. Furthermore, as Circle has announced its plan to go IPO through a SPAC transaction, which will subject the company to SEC disclosure rules, it's possible that a publicly listed stablecoin like USDC will result in better market understanding of the industry like the recent listed crypto exchange Coinbase.

While many countries have started their regulatory actions on cryptocurrency transactions and related exchanges, most likely they will only have limited impact on the decentralized businesses of major crypto players. But what happens in China and the US might be another story. What matters most to stablecoins in terms of regulation will be China and the US, the two largest crypto markets and also the two most powerful regulatory enforcers. Furthermore, US-China's new regulatory framework may influence many countries' regulatory actions against stablecoins.

But that may be a constructive development for the broad crypto market. Since stablecoins are the fundamental infrastructure for the entire digital asset industry, their complete and transparent disclosure is critical. When the space cleans up from fraud and manipulation, the broad blockchain industry can realize the true value of coins that are needed as gas for smart contracts.

Big Tech Coin: The Rise (and Fall) of Libra

Facebook's mission is to "bring the world closer together." Increasingly, that's not just about connecting friends and family to share messages, but also serving as a platform for users' financial lives. In recent years, some $100 billion in payments were enabled by Facebook annually, according to its executive running the company's financial services. But that's just the start of the social network's ambitions in the finance industry.

Since its ambitious plan on its own stablecoin "Libra" (which was renamed "Diem" under regulatory pressure), Facebook has strived to turn the world's largest social network into a financial empire (see **Figure 9.5**), and that push would only accelerate now that the company was recently rebranded "Meta," striding into the metaverse at full force.

With great fanfare, Facebook announced a plan in 2019 to create an alternative financial system based on a crypto coin and payment infrastructure. From its white paper, the mission

Figure 9.5   Stablecoin Empowers Meta Empire

is to create "a simple global currency and infrastructure that empowers billions of people." It begins with a new cryptocurrency designed for payments ranging from micropayments to remittances without fees ("as easy to send money as an email") as well as enabling more exotic "smart money" use cases like dynamic contracts, which could enable blockchain-based loans or insurance. It will provide blockchain rails for low cost, secure, and nearly immediate settlement of funds, according to the Libra white paper.

Libra, as the cryptocurrency was then called, was initially designed to be backed by a basket of developed nation sovereign currencies. The value of the coin will be pegged to a market-value basket of several trusted currencies, with individual Libras worth about a buck. It quickly drew regulatory scrutiny, because regulators worried that an independent Facebook currency creates a sovereign economy and pseudo nation state out of an internet platform. (In fact, some of the news articles even described this as "Zuck-bucks" and "Face-coin.")

Furthermore, some regulators were concerned that Libra could displace the domestic currencies of second- and third-world nations whose populations prefer a basket of first-world currencies to their own. As a result, Libra struggled to gain traction and was rebranded as Diem, which would use single-currency stablecoins denominated in the user's home currency, where cash reserves and short-dated treasuries will back each stablecoin. Still, the regulatory hurdle was too high for Diem. Ultimately, Facebook (now Meta) gave up on the project. (**See Box: "Zuck-bucks – From Libra to Diem."**)

Zuck-Bucks – From Libra to Diem

Facebook's ambitions for payments started to materialize around the time Dan Marcus, former PayPal president and then Facebook VP, announced in May 2018 that he was spearheading a new blockchain initiative inside Facebook. According to a timeline of Libra delineated by CoinDesk news report, Facebook announced seven months later that it was building a stablecoin system for WhatsApp transfers. Momentum accelerated in 2019, as Facebook moved to acquire smart contracts developer Chainspace and announced it was building a cryptocurrency that could be used inside its multiple platforms. In March 2019, it began hiring for "Blockchain Liaisons," followed by a series of actions in May 2019 including the hiring Coinbase veterans and registering "Libra Networks" as a new company in Geneva.

This wave of initiatives also attracted the wary attention of a US Senate Committee, as it issued a letter to Facebook asking for more information on plans for its crypto project in May 2019. It probably didn't help that some of the news articles described *Libra* as "Zuck-bucks" and "Face-coin."

Facebook formally announced Libra publicly in June 2019, defining Libra as part of an "ambitious vision of a decentralized, autonomous organization. . . and a borderless, easy-to-transfer means of exchange." On the day of the announcement, concerned US lawmakers called on Libra to halt its progress until more information was provided to the US government regarding its strategy and plans. In July, David Marcus agreed to testify before both the US Senate Banking and US House Financial Services committees. After a series of tussles between Facebook and the US government, Facebook stated in a disclosure document filed with the SEC at the end of July 2019 that Libra might never be properly launched.

It could be argued that Libra was doomed for many reasons, the most critical attack coming from the US and global regulators, as well as the subsequent hit from discouraged American corporate partners like Visa and Mastercard. From the US, the regulators' pushback began in June 2019, when US lawmakers raised alarm and called for an "immediate halt" to Libra's progress until lawmakers have more information on Libra's strategy and plans for the future. Policymakers kept raising concerns as Libra developed, with the US Senate Banking Committee holding a hearing on Libra in July 2019 and then US Treasury Secretary Steven Mnuchin noting that he "is not comfortable with Libra," given its possible uses for serious criminal activity.

Bearish regulators outside the US included the European Commission, which two months after Libra's public debut announced that it had begun an antitrust investigation into the workings of the Libra Association. The European Central Bank reacted in a similarly critical manner, when in September 2019, it stated that Libra could "impair the EU's monetary policy" and possibly also influence the European Central Bank's control of the Euro currency as well. In the same month, Bruno Le Maire, France's Economy and Finance minister, announced that Libra "will not be allowed to be launched in its current form."

Furthermore, because of the massive size of Facebook as a social media giant, the regulators in many countries felt that the company would run a parallel economy, especially in smaller countries, purely using the strength of its userbase, and this was not something that they were comfortable with. So, they forced Facebook to go slow on its crypto project, which by then had been renamed Diem. Notably, the People's Bank of China's head of research did not display a completely negative view of Libra, noting that Libra could accelerate PBOC's own digital currency research in the future. (As we have seen from the discussion of China's e-CNY in this Chapter, that really turned out to be the case.)

For a moment, there were signs of life for the relatively unknown Libra project when June 2019 saw Visa, alongside PayPal, Stripe, and Mastercard, according to CoinDesk reports. However, the repeated backlash between regulators and Facebook dashed the initial enthusiasm of Libra's payment collaborators. In a mere few months, by October 2019, they had all withdrawn from the Libra Association. The withdrawals left Libra with no major US payment processor, dealing a major blow to Facebook's plans for a distributed, global cryptocurrency. In the same month, Mark Zuckerberg testified before the US House Financial Services Committee, stating that Facebook will "withdraw from the Libra Association should it launch before securing all of the regulatory approvals it needs."

The year 2020 witnessed more doubts directed at Libra by regulators and heads of financial institutions around the world. Beleaguered by the lack of support from collaborating bodies, the Libra Association in April 2020 revised its white paper, announcing in a concession statement that "rather than a single stablecoin backed by a basket of assets, it will now look to issue a series of stablecoins backed by a single asset each." In other words, Diem will use single currency stablecoins denominated in the user's home currency. This would make Diem a more distributed money service provider, instead of an integrated, enormous financial system.

After the statement was issued, the latter months of 2020 saw more organizations join the Libra Association, including nonprofit Heifer International and Singapore's state-owned investment company Temasek, bringing Libra Association's total membership to 27 member organizations. In May 2020, Libra's wallet provider Calibra underwent rebranding, changing its name to Novi. Just seven months later, however, a second rebranding took place, with the Libra Association calling itself the "Diem Association," an effort CoinDesk called "part of an effort to distance itself from [Libra]'s original multi-currency basket vision."

In 2021, we saw more of a slowdown for Diem. Arguably the biggest event related to Diem was October 2021's pilot launch of Novi in US and Guatemala, where Diem was notably absent. Ultimately, Facebook (now Meta) gave up on the project. At the beginning of 2022, the intellectual property and technology behind Diem was sold for $180 million to Silvergate Capital.

But Meta's ambition to become a financial juggernaut has not ended. Now at the center of Facebook's push into payments is Novi, a digital wallet intended for users to move money around the world quickly and cheaply (free, in many cases). The Novi (formerly Calibra) digital wallet will automatically integrate to Facebook user profiles once they confirm their identity with a debit and ID card.

For more than a billion of the global population without access to in-person banking services, Novi will likely be the preferred option to send and receive payments, especially when more and more emerging market population start to have access to a mobile phone and can easily create Facebook profiles. Mobile payment applications cannot compete in terms of time and fees with Novi wallet of stablecoins, which has a giant built-in distribution platform to rely on (even if such stablecoin is not Meta's own stablecoin like libra/diem).

Similarly, we may find more Big Tech companies developing their own digital wallets to handle cryptocurrencies and stablecoins, especially when they strive to build their own metaverse based on their existing platforms in the mobile internet. For example, Twitter has decided to roll out its Tips feature to global users including bitcoin payments. The new tool gives users the ability to link to payment platforms, including the Cash App, Venmo, and others, so followers can not only show their support by liking a tweet but also by sending money.

Now the crypto industry is waiting for Elon Musk to unveil the forthcoming Web3 version of Twitter, which, as Musk puts it, is the "de facto town square" of the internet. In one of the biggest acquisitions in tech history, Musk agreed to purchase Twitter for $44 billion in April 2022 and plans to take the company private. One potential change Musk floated was making the Twitter feed algorithm "open-source," as open-source coding is the key to the decentralized ethos underpinning many blockchain applications. However, in May 2022 Musk said the Twitter deal was "temporarily on hold", while he sought more information about the proportion of fake accounts on the platform. Therefore, it remains to be seen when and if the Twitter transformation will occur soon.

As will be discussed in the following chapter, the Web3 communities have to compete not only with Big Tech companies' digital wallets and payment systems (and sovereign CDBCs), but also with Big Techs' "walled metaverses" (and sovereign digital infrastructure) by developing the creators' own "open metaverse."

CHAPTER 10

Web3 Creator Economy on Blockchain

- DAOs and Web3 Governance
- Redesign Corporations in the Metaverse
- YGG and Open Metaverse vs. Big Tech Platforms
- Challenge 1: Interoperability and the "Internet of Blockchains"
- Challenge 2: State-backed Metaverse
- Challenge 3: Will Web3 startups become new "Big Techs"?
- Creator-Verse: Future Creator Economy

Constitution DAO and Web3 Governance

In November 2021, thousands of individuals from around the world came together as a decentralized autonomous organization (DAO) to bid on 1 of 13 extant original copies of the US Constitution. (DAOs are self-managed organizations that are defined by a transparent set of rules encoded as a computer program. The members of a DAO have their incentives aligned through such mutually agreed upon governance and, often, through the use of a specific cryptocurrency or token.)

In exchange for contributions, the "ConstitutionDAO" participants received digital tokens that carried the right to vote on decisions about the document. Supporters rallied around the idea of using a democratic decision-making body to take ownership of one of the world's most storied symbols of democracy. The DAO crowdfunded approximately \$47 million in cryptocurrency, but eventually lost the auction to Ken Griffin, the CEO of the hedge fund Citadel. Nevertheless, it showed that people with shared beliefs could organize to compete in establishment forums previously available only to the wealthiest individuals. (See **Box: Constitution DAO.**)

ConstitutionDAO

In some ways, ConstitutionDAO is reminiscent of the January 2021 GameStop short squeeze, organized by SubReddit r/wallstreetbets, due to its decentralized organization style and financial momentum gained in a short period of time. Although ConstitutionDAO lost out on its bid to Ken Griffin, founder of hedge fund Citadel, it was still an astonishing display of how fast a DAO could mobilize and deploy massive funds for a shared cause. It is also worth noting that Sotheby's, the auction house that the US Constitution copy was sold through, had never previously dealt with a DAO.

According to a report from The Verge, ConstitutionDAO was founded for laughs. It seemed to the group's members that it would be funny for a group of random people from the internet to buy something as holy and serious as a US constitution document. What then began as a joke on Discord spun off into a gigantic fundraising event on Ethereum.

A 2021 TechCrunch article reported that within a week of the ConstitutionDAO's Discord launching, the DAO had raised over \$40 million worth of ETH on the DAO platform Juicebox. The Discord server of ConstitutionDAO had ballooned from 0 to over 8000 members, and ConstitutionDAO's official Twitter account estimated at the end of the event: "We had 17,437 donors, with a median donation size of \$206.26. A significant percentage of these donations came from wallets that were initialized for the first time." According to the aforementioned *The Verge* report, if ConstitutionDAO had succeeded in bidding for the copy, the DAO planned to decide the document's future by voting, weighted by how many governance tokens each contributor had.

Even though ConstitutionDAO had failed to win the US Constitution copy bid, it still is a mesmerizing display of what promises DAO and the greater concept of Web3.0 will have for us all. If we take ConstitutionDAO as a case in point, while a traditional corporation or venture capital firm would have been able to raise funds, it might not have been able to pull it off with the same jaw-dropping speed or garner the same type of media attention for its moves, as corporations and venture capital firms are, by default, usually more centralized and slower in decision-making.

While the ConstitutionDAO was ultimately outbid, the movement propelled DAOs into mainstream news headlines and inspired a wave of similar projects, such as DAOs that have announced plans to buy a golf course (LinksDAO), an NBA team (the Krause House DAO), and land in Wyoming (CityDAO). Some may see these half-baked undertakings as evidence of the crypto economy's recklessness, but these spicy stories only present a partial picture of the burgeoning DAO phenomenon. DAO is more than crowdfunding to buy fancy stuff; since decentralization, transparency, and shared ownership are the cornerstones of Web3, perhaps DAO structures will revolutionize how human communities organize going forward.

DAOs have been formed for various purposes (see **Figure 10.1**). The current DAOs developed by smart contract platforms mainly include agreement DAO, investment DAO, grant DAO, service DAO, media DAO, social DAO, and collection DAO. For example, many DeFi applications, such as Uniswap, Aave, and MakerDAO, are governed by DAOs, which provide a mechanism for protocol development and treasury management through self-executing smart contracts on

Leading DAOs by Category

Figure 10.1   DAOs Formed for Various Purposes
Source: Coinbase, "DAOs: Social networks that can rewire the world," 2021

the blockchain. Compared to traditional asset management business, the DAO has no need of a trusted investment manager; instead, its cycles of operation are embedded in smart contracts on the blockchain.

So, the bigger question is: The corporation was the default mode of organizing human activity in the private sector over multiple centuries, but what should be the optimal governance model on Web3, when the nature of the economic puzzles that corporations evolved to solve have shifted? DAOs enable individuals to collaborate, manage projects, and own and invest assets like a traditional organization, but they can provide far greater levels of transparency, openness, and democratic governance. DAOs may become the default mechanism for facilitating collaboration in the Metaverse.

Redesign Corporations in the Metaverse

As we compare DAOs to corporations, it's worthwhile to note that most corporations in the world today are structured in essentially the same way as they were in the 1600s: companies (formed and run by the management team) accept funds from investors, who become the shareholders of the companies, in exchange for the responsibility of maximizing shareholder value as their primary (and sometimes only) mission.

This corporate structure has served as the fuel of capitalism – both the good and the bad. The good has led to the modern capital markets and significant improvements in business efficiency, while the bad has frequently manifested itself in the pursuit of profits above all else, even at the cost of ESG (environmental, social, and governance) considerations. Furthermore, because of the centralized governance structure, the companies' executives and large shareholders can often hold a disproportionate degree of control.

For example, in the current internet economy, the creators, whether the writers on Twitter or the short video-makers on TikTok, have nearly zero influence on how those platforms operate and have little claim on the revenues generated by the contents they create. For the billions of users, the monopoly of data by giants has been concerned by the society.

Because of their unique structure, DAOs offer the promise of enabling a focus on community, rather than just profit, and might offer a more socially conscious structure. DAOs have the potential to improve upon a style of corporate organization that hasn't meaningfully changed in hundreds of years, enabling creators and users to participate in global collaboration more widely in the digital world (see **Figure 10.2**).

Why could DAOs be the governance for the Metaverse? The short answer is that blockchain and crypto assets have started as a DAO. Recall the discussion at the start of cryptocurrency in **Chapter 3**, and you may realize that the Bitcoin network is the first and simplest DAO. Anyone can join the Bitcoin network at any time to become a node, providing computing power to earn bitcoins and ensure the security of the ledger at the same time. Every time you buy some bitcoin, for example, that transaction gets recorded to the Bitcoin blockchain, which means the record is distributed to thousands of individual computers around the world. Equally important, you may easily leave the ecosystem at any time.

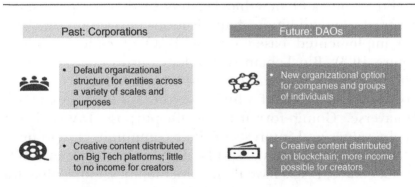

Past vs. Future: Corporations vs. DAOs

Past: Corporations		Future: DAOs	
	• Default organizational structure for entities across a variety of scales and purposes		• New organizational option for companies and groups of individuals
	• Creative content distributed on Big Tech platforms; little to no income for creators		• Creative content distributed on blockchain; more income possible for creators

Figure 10.2   Corporations vs. DAOs in Creator Economy

As a public blockchain, Bitcoin is also transparent. All transactions are available for anyone on the internet to see, in contrast to traditional corporate bookings. (Many people thought that crypto, especially bitcoin, is untraceable and is primarily used for nefarious purposes, but that's a huge misconception. The blockchain is a public ledger of activity. It's therefore possible to track the movements of funds from one account to another.) This decentralized recording system is very difficult to control or manipulate.

From the example of Bitcoin, we may see some of the unique advantages of DAO (compared to corporations), such as:

- **Maximum automation and minimal costs for ecosystem administration.** No employees are needed from the administration perspective except for the mere existence of a DAO.
- **Flexibility.** In terms of organization structure to create roles, contributors may join and leave freely contributors from the community and outside contractors can work from all over the world, not necessarily restricted to one or a particular location.
- **Transparency.** All types of transactions are traceable and auditable by all permitted parties, resulting in much higher transparency and fraud reduction.

Ethereum and new blockchains further support smart contracts, and all kinds of applications derived on this basis are implemented based on the DAO of code rules (see **Figure 10.3**). Blockchain technology ensures "code is law," while DAO facilitates the orderly formulation and execution of rules, both of which are the cornerstones of the Metaverse. Going forward, specific-purpose DAOs driven by the values and interests of their community are going to become more common, as will collaboration between DAOs.

Of course, DAOs have their vulnerabilities, too. Unlike traditional entities, there is a lack of legal personality or central

Democratizing Ownership of Organization through Smart Contract on Blockchain

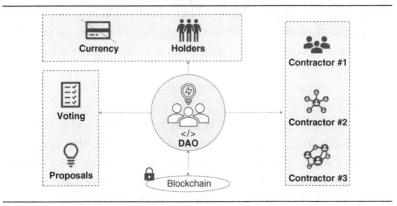

Figure 10.3   Smart Contracts Expand the Implementation of DAOs

Source: Adapted from CodeCentric.de Blog, 2017

responsibility of DAO. While corporations have the CEO and board chairman, from a legal perspective DAO basically means "all of its participants." For that reason, it's not clear who should represent the DAO in a judicial proceeding and file a tax return for purposes of tax law. In the end, it is difficult to identify the entity or a single person responsible for DAO's operation.

Most importantly, the key aspect of its structure – the smart contracts – that makes a DAO nimble and autonomous in the virtual world also creates challenges in the real world. The main assumption of DAOs is that the rules for operation of a DAO are solely based on the code of smart contracts on a blockchain. However, our actual lives don't take place solely in the metaverse. We are still very much in the physical world. As such, the DAO structure will have to take into account the important connections to real-world activity, which have not yet been completely tested in existing DAO examples.

For example, if the ConstitutionDAO successfully won the bid for the rare copy of the US Constitution, it may need to engage with a lawyer, custodian, and insurer, which are all individuals and institutions in the physical world. (They are the "contractors" in **Figure 10.3**.) Therefore, DAOs must establish some method of dealing with real-world relations, protections, benefits, and liabilities as traditional corporations do. These tough challenges must be addressed before DAOs can realize their full potential to transform the real world (in addition to the virtual world).

To summarize, DAOs are very much uncharted territory for now. To fully realize their potential benefits in the real world, DAOs have some significant hurdles to overcome. But the DAO governance is poised to replace the arcane corporate system to develop a new Web3 system, where the user will have more control of their data, the individual creators can monetize their content more directly with their fans, and more flexible payment mechanisms will be available for all participants. For a case study, in the next section we will examine how the gaming industry will evolve with more decentralized governance. (As discussed in **Chapter 6**, gaming is an early version of the Metaverse.)

YGG and Open Metaverse vs. Big Tech Platforms

YGG is an excellent example of a robust DAO on blockchain. YGG is a key player in the play-to-earn (P2E) gaming ecosystem that is often mentioned alongside Axie Infinity (see the introduction in **Chapter 6**). According to YGG's official website, YGG is a global P2E gaming community where players can earn real money by playing NFT games like Axie Infinity, The Sandbox, League of Kingdoms, and other blockchain-based games. The company's 2021 white paper defines YGG as a DAO for investing in NFTs used in virtual worlds and blockchain-based games. YGG's mission is to create the world's biggest virtual economy, optimizing its community-owned assets for maximum utility and sharing its profits with token holders.

YGG, as hinted at in its name "Yield Guild Games," is formed as a DAO and shares traits with medieval guilds. The YGG game players, like tradespeople in the earlier professional guilds, benefit from the association effect of being in a united organization. The YGG Guild provides for individual players (i.e., "YGG scholars") a means to rent NFTs (such as Axies) to begin playing and earning, similarly to how tradespeople in medieval guilds can request tools or economic assistance from guild members to get their own professions off the ground. YGG's association effect benefits its guild members, as each member might find it difficult to reach such a variety of games or fellow players by themselves individually, versus reaching them in the YGG ecosystem with relative ease. (See **Box: Medieval Guilds and YGG Scholars.)**

As YGG is a DAO, it has features of a DAO-type organization where members of YGG can vote on community issues, such as the types of games that should be playable and the kinds of virtual assets to invest in. Gabby Dizon, co-founder of YGG, stated in a report in *BusinessWorld*: "At its core, YGG is a community of P2E gamers. Think of it as a massively multiplayer online (MMO) guild, for example, but operating across several games, investing in yield-generating NFTs within those games, and lending those in-game assets and inventory out to our player base." In the post-Covid era, YGG has provided for its members a metaverse platform to meet, organize, earn, and mobilize, as a guild should rightfully do.

Medieval Guilds and YGG Scholars

What is truly intriguing about YGG is hinted at in its name: YGG is a metaverse-based guild. In medieval times, artisans of diversified trades would build guilds in the interest of professional networking, economic cooperation, and mutual protection (in matters of both personal safety and market participation). Guilds as a form of business organization have been around since the Middle Ages, and have survived as a professional organization form in modern times as the guilds for artists and writers.

YGG parallels medieval guilds most prominently in the economic cooperation and mutual benefit element. It began out of an act of generosity. In
2018, Gabby Dizon, a veteran in the gaming industry, lent his Axie Infinity token
"Axies" to other players who could not afford their own Axies (players who want
to earn via Axie Infinity need to acquire three Axies, see related discussion in
Chapter 6).

When Dizon realized how his fellow Filipinos leveraged Axies as a source of
income during the Covid-19 pandemic, he decided to create a global NFT gaming community that could drive more positive economic growth for players and
investors alike. Dizon, according to YGG's official website, then founded YGG
with fintech veteran entrepreneur Beryl Li and a developer under the pseudonym Owl of Moistness.

With its "YGG Scholarship" program, YGG rents Axies to its "scholars,"
enabling the scholars to play games with no upfront cost to the scholar. Scholar-
ships enable players and the YGG platform to share revenue, with the scholars
playing the game taking the largest proportion of their gameplay earnings, and
YGG taking a smaller cut of the scholars' gameplay earnings to fund itself as
a platform. YGG's official website states that some scholarship programs also
offer training and mentorship benefits, so that players can improve their gaming
strategy and become better at gaming, earning more for themselves and YGG.

Of course, the Big Techs of existing internet economy will
not sit idle when gaming DAOs like YGG expand. They also
see gaming as a promising path to the Metaverse, and they
are moving aggressively into the video game industry. In 2022,
internet and entertainment giants like Microsoft, Sony, Take
Two, Tencent, and Warner Music Group have announced
unprecedented M&A and partnership transactions with gaming companies (see **Table 10.1**). This ongoing string of major
acquisitions, already exceeding $70 billion, will accelerate the
Big Tech's stride into blockchain gaming and shake out the
gaming landscape completely.

Among them, the Microsoft/Activision Blizzard acquisition
is the most remarkable. In early 2022, Microsoft announced
buying Activision Blizzard in a $68.7 billion all-cash acquisition.
It's the software maker's biggest deal ever, almost three times as
large as the 2016 purchase of LinkedIn. The transaction, if it
can get regulatory approval, will create the world's No. 3 global

Table 10.1   Unprecedented Gaming M&A and Partnership Transactions in 2022

Buyer	Target	Price for Acquisition	Blockchain Gaming Initiative
Microsoft	Activision Blizzard	$67 billion	With more resources from Activision, Microsoft will accelerate its NFT gaming business
Take Two	Zynga	$12.7 Billion	Zinga formed a partnership with blockchain gaming powerhouse Forte in December 2021
Sony	Bungie	$3.6 billion	Bungie is exploring blockchain gaming as a developer
Tencent	Sumo Group	$1.27 billion	Sumo partnered with the blockchain-company Dapper Labs and explored gaming on Flow blockchain since 2020
Warner Music Group	Sandbox	Undisclosed	Sandbox is a P2E gaming platform

gaming company in Microsoft, just behind China's Tencent, the world's largest online entertainment company (publisher of League of Legends, holding 40 percent of Epic Games and 100 percent of Riot Games) and PlayStation maker Sony.

What's in the background is that Microsoft has been an active player in gaming for a while. For many, Microsoft is known as a software company. In recent years, however, Microsoft corporate strategy has been coalescing around cloud, content, and creators. For example, it is responsible for developing and publishing some of the biggest franchises in history: Age of Empires, Forza, Gears of War, Halo, Minecraft, Fallout, Microsoft Solitaire, Microsoft Flight Simulator, DOOM, The Elder Scrolls, and many more.

Gaming is one of Microsoft's two big metaverse plays, and Microsoft has already worked on NFT gaming before this massive acquisition. (The other is Office and conference software. Microsoft is integrating virtual-reality offices and 3D avatars into its Teams remote-collaboration software.) The acquisition will offer even more devoted game communities to create their own gaming metaverses. The goal of Microsoft is to create a gaming empire big enough that gamers will come to it directly, bypassing Apple. (Microsoft has been at war with Apple and Google over the fees the app stores charge for games.)

Yet, there is something fundamentally at odds when a platform, like Microsoft or Meta, whose primary business model is an intermediated marketplace founded on centralized data, yet pivoting to build something aiming to be open, interoperable, and owned by users. The biggest difference between Big Techs' metaverses and DAOs' metaverses, is that the latter operates on open, permissionless, blockchain architecture – any developer can come and build a metaverse application on an open blockchain, and any user can acquire and trade their own virtual assets. That's the vision of the "open metaverse."

However, before this vision can materalize, the open metaverse has three major challenges to overcome, which will be discussed in the following sections.

Challenge 1: Interoperability and the "Internet of Blockchains"

One important aspect of the "open metaverse" evolution is the inter-connectivity of such a computing landscape, which may lead to the erosion of the walled garden elements of the mobile computing age. The reality is that, while blockchain's potential for improving business processes and creating new business models, providing transactional transparency and data security in the value chain, and reducing operational costs for data management and exchange is obvious for many individuals and corporations, the expected mass adoption failed to happen till now. What has been holding the blockchain? A lack of uniform technical standards and interoperability.

The main challenge is that numerous promising blockchains are growing in parallel because the traditional architecture of blockchains does not allow them to communicate with each other. The individual blockchain networks are not inherently open, and they have different characteristics – such as the type of transactions, hashing algorithms, or consensus models. Every blockchain network represents an entirely new set of records and hosts different applications. This is true for every major blockchain, including the two leading chains of Bitcoin and Ethereum. This has resulted in a series of unconnected blockchain ecosystems operating alongside but siloed from each other, each supported by a weak network of nodes and susceptible to attack, manipulation, and centralization.

Therefore, it's critical for the Web3 to develop solutions that help these blockchain networks interoperate. This ability of blockchain networks to communicate and share data with each other is referred to as blockchain interoperability. The interoperability of blockchain today can be compared to the cell phone and email interoperability at the beginning of the internet.

Imagine – Emailing would be useless if two email platforms built on two different infrastructures were not interoperable. For example, what if you couldn't send an email from your Gmail account to an account on Microsoft Outlook? The same is the case with mobile and computer operating systems. What if you couldn't call an Android smartphone user from your iPhone? Or, what if two users using Zoom on Windows and macOS couldn't video meet each other? The lack of interoperability would have made the internet extremely difficult to use.

The interoperability not only means the possibility that disparate blockchain systems can communicate with each other. Above all it is the ability to share, see, and access information across different blockchain networks without the need for an intermediary – like a centralized exchange. As such, a growing number of interoperability projects have entered the scene to try to bridge the gap between the various blockchains. The vision of interoperable blockchains thereby rests on several functionalities and abilities including integration with existing systems, transactions across multiple chains, and empowering apps and making it easy to switch one underlying platform for another.

The most efficient and scalable way to build interoperability is through the joint effort of establishing industry standards as well as identifying a "network of networks" structure that industry networks can converge around. Such a blockchain network represents a "web" of interconnected networks. This architecture would allow an organization to connect and transact with multiple solutions, not restricted to a single network, and open up a market of interoperability across solutions. The network of networks model for interoperability continues to gain momentum, and the crypto world starts to see promising blockchain hubs emerge.

For example, Cosmos is one of the most prominent interoperability solutions, which aims to act as an ecosystem of blockchains that can scale and interoperate with each other. Cosmos is a smart contract platform that has prioritized interoperability for its blockchain design. The architecture is based on the "hub-and-spokes" system whereby a series of "spoke" chains connect to a "central" hub by means of inter-blockchain communication. The Cosmos goal is to create an "internet of blockchains" – a network of blockchains that can communicate with each other in a decentralized way.

In April 2021, Cosmos hit a historic milestone by launching the Inter-Blockchain Communication protocol (IBC), the Cosmos standard for blockchain interoperability. IBC enables independent blockchains to connect, transact, exchange tokens and other data, scale, and thrive in an interconnected network. In just eight months, we have witnessed the growth of a robust new economy. By the end of 2021, 25 chains had officially joined the IBC. They have all been able to connect to various decentralized exchanges (DEXs) available in the Cosmos ecosystem, such as Osmosis, Gravity DEX.

To solve the interoperability issue, Cosmos proposes a modular architecture with two classes of blockchain in the IBC: Hubs and Zones (see **Figure 10.4**). Zones are regular heterogeneous blockchains and Hubs are blockchains specifically designed to connect Zones together. When a Zone creates an

Cosmos IBC Hubs and Zones

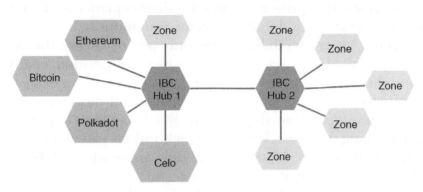

Figure 10.4   Internet of Blockchains – Cosmos IBC Hubs and Zones
Source: Cosmos

IBC connection with a Hub, it can automatically access (i.e., send to and receive from) every other Zone that is connected to it. As a result, each Zone only needs to establish a limited number of connections with a restricted set of Hubs. Hubs also prevent double spending among Zones. This means that when a Zone receives a token from a Hub, it only needs to trust the origin Zone of this token and the Hub.

What is promising is that IBC does not stop at Cosmos-built blockchains. In 2022, Cosmos was working on connecting to many other major ecosystems including Bitcoin, Ethereum, Polkadot, and Celo, potentially unlocking vast amounts of liquidity that will flow through this "internet of blockchains." Meanwhile, more interoperability projects like Cosmos are also building up similar ecosystems of blockchains that can scale and interoperate with each other.

Going forward, the arrival of interoperability solutions may fundamentally change present attitudes toward blockchain. Interoperability is crucial in any software system – it simply will not reach its full potential if it can't work with other software. In the past, computer scientists spent decades constructing networks we now know as the Internet, and the Internet only became transformative once American-developed TCP/ IP programming protocols gained wide acceptance as the way to make a variety of different systems communicate.

Today, interoperability is the only way to realize the full promise of Metaverse, where interoperability would enable smoother data and solution sharing, easier execution of smart contracts, more user-friendly experience, and more opportunity to develop partnerships. It is likely to become an important game changer for blockchain to reach widespread acceptance and adoption. The Metaverse will only arrive when all blockchains can be integrated into one seamless internet of blockchains to challenge the existing Big Tech platforms.

Challenge 2: State-Backed Metaverse

Just like in the context of digital currency, where a three-way competition among the cryptocurrencies, Big Tech tokens, and CBDCs (central bank digital currency), relating to Web3 blockchain infrastructure, the open metaverse must compete with both Big Tech platforms (see the earlier discussion of DAO v. Corporation) and government-driven metaverse. As discussed in **Chapter 9**, China's treatment of cryptocurrencies, NFTs, and blockchain technology fits into a broader pattern of the government's effort to build an "internet with Chinese characteristics," and the same is true for the metaverse infrastructure. (Reference to Winston Ma's 2021 book *The Digital War: How China's Tech Power Shapes the Future of AI, Blockchain and Cyberspace.*)

Similar to promoting blockchain technology (including developing China's sovereign digital currency e-CNY) and restricting crypto trading at the same time, China made a distinction between metaverse technology and metaverse tokens. In December 2021, Shanghai, China's most populous city, announced that metaverse R&D would be an integral component of the city's five-year development plan. The city's government will aim to leverage the metaverse "in areas such as public services, business offices, social entertainment, industrial manufacturing, production safety and electronic games." The governments of the cities of Beijing, Wuhan, and Anhui, as well as the provinces of Zhejiang and Hefei, have since followed Shanghai's example.

Meanwhile, entering the metaverse has been all the rage for Chinese internet companies. Major companies like Tencent, Alibaba, and TikTok owner ByteDance have all recently announced plans to begin developing technologies that will potentially make them key players in the Metaverse, which is critical at the time when their business models on smartphones and mobile internet have matured. Since 2021, the number of registered trademarks in China related to the Metaverse has skyrocketed (see **Figure 10.5**).

However, with public blockchains like Bitcoin and Ethereum illegal in China, the country is building a new version of metaverse infrastructure on its state-backed Blockchain Services Network (BSN), which runs what amounts to centralized, private, permissioned blockchains. Such infrastructure is open to developers, so the companies operating on them

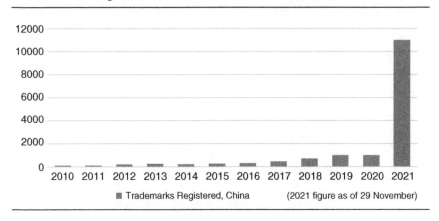

Number of Registered Trademarks in China Related to the Metaverse

Figure 10.5   Metaverse-related Trademarks Mushroomed in China
Source: Quartz

may collect users' identity data for regulatory purposes. For metaverse applications that rely on cryptocurrencies, China's digital currency might be used instead in metaverse payment applications. (See **Box: The Open Blockchain-based Service Network** for background on BSN.)

The Open Blockchain-based Service Network (BSN)

The Open Blockchain-based Service Network (BSN) is jointly developed by central government, state corporations, and private tech companies (see **Figure 10.6**), including the Chinese government policy think tank State Information Center (SIC), which is under NDRC (National Development and Reform Commission, which also set up an AI national development policy), China's state-run telecom giant China Mobile, the Chinese government-supported payment card network China UnionPay, and the private tech company Red Date Technology.

At the April 2020 launch, the Blockchain Service Network Development Alliance stated that the network had 128 public nodes. China had 76 of these nodes already in the network at that time with 44 under construction, and the remaining 8 were overseas city nodes (including Singapore, which already tested BSN during its pilot phase), which covered six continents. Therefore, at its inception and within China alone, the BSN ecosystem was instantly the largest blockchain ecosystem in the world.

In 2021, the BSN began to venture beyond China. In September 2021, the BSN launched its first international expansion portal in South Korea, managed by Red Date Technology and in partnership with Korean blockchain company MetaverseSociety Corp. According to Forkast news, MetaverseSociety will exclusively operate a BSN portal offering blockchain-as-a-service in Korea, enabling blockchain developers in Korea to be able to develop blockchain

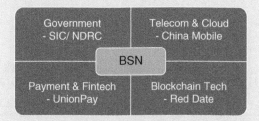

Figure 10.6   BSN – The State-Backed Network

applications, integrating frameworks such as Ethereum, EOS, Polkadot, NEO, Tezos, and Oasis.

Besides South Korea, BSN efforts, taken by Red Date Technology, have also ventured to Singapore. In July 2021, CoinDesk reported that Red Date had registered a nonprofit foundation in Singapore to manage the international version of the BSN. The hope in Singapore for Red Date is to make the BSN into an "international standard open-source community."

BSN has also expanded its partnership efforts outside of Asia. In October 2021, Red Date Technology had set up two new BSN portals in Turkey and Uzbekistan, signing an agreement with a Turkish consultancy firm, Turkish Chinese Business Matching Center (TUCEM). CoinDesk reported that the new portals will allow blockchain developers in Turkey and Uzbekistan to build BaaS applications using the global BSN portal hosting major blockchains like the Ethereum network, Algorand, Polkadot, Tezos, ConsenSys Quorum, Corda and others.

In summary, the BSN is set to include as many blockchain frameworks as possible and make them accessible under one uniform standard so that it could be deployed nationally and globally at a low cost. In China, all city governments, state-owned enterprises, and IT framework operators are gearing up to adopt and interoperate with the protocol. Its multi-cloud architecture has already included China Unicom, China Telecom, China Mobile, and Baidu Cloud from the domestic sector, as well as Amazon AWS, Microsoft Azure, and Google Cloud of the West.

According to its official white paper, BSN is a cross-cloud, cross-portal, cross-framework global infrastructure network used to deploy and operate all types of blockchain applications. Previously, each participant in a traditional consortium chain application must build and operate its own exclusive node and respective consensus mechanism. Each node must use a physical server or cloud service to connect with one another through the internet or an internal network, thereby forming an isolated blockchain application similar to a local-area network.

In this isolated blockchain structure, the application designer needs to establish a new blockchain operating environment for each consortium chain, which is highly costly. For example, to deploy blockchain applications at the platforms of major cloud providers (such as Alibaba, Huawei) may cost the users tens of thousands of dollars a year. Worse, the server resources are often not fully used. According to Red Date, the tech firm behind the BSN, only 2–3 percent of enterprise users would need more than 1000 transactions per second (TPS), which allows users to make full use of the cloud services. As a result, the high cost brought by deployment and maintenance is a major barrier for blockchain entrepreneurs.

The BSN can be attractive to the developers of blockchain applications, because:

1. **The BSN launch will allow companies to access ultra-low cost blockchain cloud computing services because it enables customers to have the services in much smaller units.** Users can pay for exactly how much they need. Target pricing is less than US\$400 per year, which would allow any SME or individual access to the critical tools to participate in the digital economy and drive adoption and financial inclusion opportunities.

2. **The BSN aims to simplify blockchain application developments.** It provides a public blockchain resource environment for developers with the concept of the internet. Just like building a simple website on the internet, developers can deploy and operate blockchain and distributed ledger applications conveniently and at a low cost. According to Red Date, building an application for 80 percent of the BSN users would be as easy as filling out a form online. ("They don't even have to write their own smart contracts; all they need to do is select one of them in our system.")

3. **The BSN plans to enhance the connectivity between different blockchain applications.** In the traditional isolated blockchain context, various applications of different technical standards cannot be unified, thus business data is unable to interact with each other, which restricts the broad adoption of blockchain technology (see **Figure 10.7**). The BSN white paper indicates developers

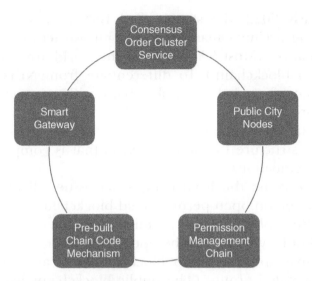

Figure 10.7   Five Major Parts of BSN Framework

can use a single private key to deploy and manage applications on multiple frameworks, and at the same time, to realize interconnectivity and mutual communication between them. (Within this process, each framework retains the unique features of its own smart contract and consensus mechanism.)

The BSN network has built up a substantial blockchain economy, and in 2022 it expands its infrastructure to support the NFT plays. Like in the US, the NFTs market has grown rapidly in China since the second half of 2021, and Google search data indicates that Chinese consumers have a massive appetite for information on NFTs, with searches for the term being made much more frequently in China than in any of the other 70 countries surveyed. But rather than ban NFT issuance outright (like cryptocurrencies), China attempts to control the market by creating its own NFT infrastructure that requires users to verify their identities and allows for state intervention where regulators suspect illegal activity is taking place (e.g., money laundering and investment fraud).

In early 2022, the state-backed BSN announced it was developing a China market-specific infrastructure to enable NFT issuance. (Most NFTs around the world are part of the Ethereum blockchain.) To differentiate from NFTs outside China that are traded on public chains with cryptocurrencies (which are banned in China), NFT is renamed as Decentralized Digital Certificate (DDC) by BSN, and the BSN-DDC network is a structure for building NFTs that is compliant with Chinese regulations.

To overcome the legal compliance issue, all the official DDCs will be on open permissioned blockchains, which combine features of public and permissioned chains. BSN turned to a technology known as the open permissioned blockchain (OPB), an adapted version of blockchain that can be governed by a designated group. (The public blockchains like Bitcoin and Ethereum are "permissionless," for which a node can join and leave freely.)

According to BSN, it has already integrated 10 Open Permissioned Blockchains available on the BSN-DDC. These are localized versions of their permissionless counterparts that set restrictions on who can participate in network governance and use fiat currency for payment. The 10 OPBs include adapted versions of the Ethereum (Ethereum-based Wuhan Chain) and Corda blockchains (Corda-based Zunyi Chain), as well as Wenchang Chain powered by Cosmos-based IRISnet, EOS-based Zhongyi Chain, and domestic blockchains such as FISCO BCOS-based Tai'an Chain, initiated by Tencent-backed fintech firm WeBank.

Essentially, DDCs are the same as NFTs, but renamed to emphasize their uses for certification. The NFT/DDC technology is a digital certification and distributed database technology that can be applied in any scenario where digital proof is required. While NFTs are currently used mostly for authenticating digital artworks, in China the biggest market lies in certificate and account management for all types of businesses. Take car number plates, for example. Such a DDC system would give the car owner, government, and insurer controlled access to data such as mileage, engine number, and repair history, with each party being aware of the others' rights.

BSN said the BSN-DDC infrastructure would offer "a diverse, transparent, credible, and reliable" one-stop shop for businesses to mint and manage their own NFTs without relying on cryptocurrencies, which are banned in China. All payments on the network will be made in fiat currency via traditional noncrypto means such as bank cards, Alipay, or WeChat Pay to comply with local regulations. Similar to the low cloud cost for DApp developers on BSN, the minting fees for NFTs can be as low as 5 cents in Chinese CNY (0.7 US cents), which is significantly lower than public chains.

Such cheap and convenient BSN infrastructure may attract developers and users, especially for those who are more focused on compliance issues and less concerned about "permissionless" and "decentralization" (see **Figure 10.8**). As such, the BSN and BSN-DDC networks represent the rise of a state-backed metaverse ecosystem, which may expand quickly as China (and more countries) accelerate their regulation of the Metaverse.

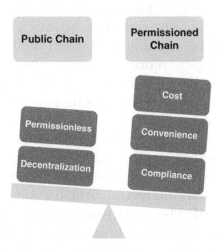

Figure 10.8   State-backed Metaverse Infrastructure has "3C" Advantages

Challenge 3: Will Web3 Startups Become New "Big Techs"?

The vision of open metaverse is decentralized ownership and shared interests, which is in direct contrast with the centralized big tech platforms in the current digital economy. While we are still years away from the ultimate Metaverse, we have started to see some progress being made on opening up walled gardens. Epic v. Apple is a great example.

Game platforms took on Big Tech in 2021 as Epic Games sued Apple for antitrust violations. The case is an important conflict between a platform owner (Apple) and a major game company (Epic Games) that could set the rules of engagement and competition in an era where gaming is the biggest media and social network. Epic Games challenged Apple's policy of collecting a 30 percent fee on every in-game transaction in titles like Fortnite. The federal judge on the case ruled that Apple violated California's laws against unfair competition. Still, she ruled in favor of Apple on other important counts in the complicated antitrust lawsuit.

While Epic largely lost almost all of the charges it levied against Apple in its antitrust case, it did win on one important point: the right to promote in-app alternative payment methods off the Apple store. The appeals court has stayed this victory while the litigation continues, but it's a crack in Apple's empire. This ruling could give alternative payment providers and game developers more hope of capturing the revenues they generate, as they will be at least able to promote lower prices for digital goods on websites that are off the app stores.

Epic vs. Apple underscores the concentration power of modern content platforms, which hold game developers captive. The case victory may be enough to spur discord among game and app developers against the status quo of app stores. On top of that, Epic still has an antitrust suit pending against Google over Google Play Store practices, which is still developing in 2022. Epic's aim, according to the company's executives, is to turn "Fortnite" into a platform on which independent developers could distribute their games and other forms of entertainment online and earn more of the profits themselves.

What's ironic is that gaming giants like Epic Games (Fortnite) and Roblox may become the new Big Tech companies, even though the Web3 vision is that no mega corporation would be dominant. Quite likely, the gaming giants could one day boast dominance in the 3D, interactive internet like that Apple, Google, and Big Tech enjoy today in the existing two-dimensional internet. More importantly, even for the more "pure" Web3 startups that focus on "decentralized" services, the leading players have accumulated monopoly power by owning the rails in a familiarly centralized way (see **Table 10.2**).

For example, OpenSea, the world's largest NFT market-place, is probably more concentrated than any other exchange platform in the capital markets. OpenSea has captured more than 90 percent of the NFT trading market traffic globally, and the marketplace takes a 2.5 percent cut of every transaction (much higher than stock trading on "concentrated" Nasdaq stock exchange). There are many NFT trading platforms in the market (such as SuperRare), but OpenSea, which controls huge traffic, is beyond the reach of other platforms. (See background information of OpenSea in **Chapter 5.**) In the pursuit of decentralization, OpenSea is a dominant existence in the NFT sector.

Table 10.2   Monopoly Power by Web3 Dominant Players

Dominant Players	Crypto Sectors	Monopoly Index
Gnosis Safe	Multi-signature management of crypto assets	99%
OpenSea	NFT marketplace	90%
Chainalysis	Cryptosecurity and compliance	90%
Chainlink	Oracle	90%
Metamask	Crypto wallet via browser extension	85%
Uniswap	DeFi exchange	80%
Bitmain	Crypto mining equipment	70%

Source: Adapted from Wu Blockchain, March 2022 report

As you would have imagined, DAOs came up to decentralize the OpenSea giant. The year 2021 was "the year of NFTs," and the biggest NFT event to end 2021 was the $SOS token airdrop by OpenDAO to the users of OpenSea, the largest NFT marketplace in terms of users, trading volume, and number of artworks for sale. The catalyst was likely the rumor about a potential IPO of OpenSea, which will turn it into a government-registered corporation (but all the NFT creators on the platform have no share in the pie). Thus, OpenDAO was created as a digital native community DAO, centered on leveling the playing field for both creators and collectors. (See **Box: OpenDAO Airdrops SOS on OpenSea.**)

The airdrop introduction from OpenDAO described the campaign like this: $SOS thanks all NFT creators, collectors, and markets for nurturing the entire NFT ecosystem. Special thanks to OpenSea for its leadership in promoting NFT trades. To pay tribute, we chose OpenSea collectors for the airdrop. The key message, however, was that all NFT contributors deserve to be rewarded. Essentially, OpenDAO tried to create a real "creator-driven" platform based on the users of OpenSea, which was viewed as a "centralized" platform. It represents a battle between Web2.0 and Web3.

OpenDAO Airdrops SOS on OpenSea

Some of Web3 business promotion moves are almost a lift-and-drop versus those in the earlier internet economy. But the marketing tactics they use and organization structures they deploy are still similar. The 2021 SOS Airdrop and its affiliation with OpenSea is a great example.

According to a December 2021 report from Decrypt, Christmas Eve 2021 witnessed an airdrop surprise for anyone who had ever spent any money on OpenSea on NFTs, as they could claim a free Ethereum token called SOS (CRYPTO: SOS). SOS is an airdrop token from a DAO named OpenDAO, and while the project is not officially affiliated with OpenSea, close to 200,000 OpenSea wallets had claimed SOS tokens by December 2021. In just two days after the airdrop, the SOS token shot up 1000+ percent, and the contract behind SOS saw a market capitalization of $200 million.

According to news site Benzinga, the OpenDAO website had offered three simple steps for OpenSea users to claim their SOS tokens: (1) connect Open-Sea wallets, so that data on how many NFTs (and money) were transacted on OpenSea could be read; (2) estimate OpenSea users' reward; and (3) initiate the claim for SOS tokens, for which OpenSea users have until June 30, 2022, to claim (unclaimed SOS will be sent to the DAO Treasury system). Users can then add the SOS tokens claimed to their wallets.

While the startling growth in SOS' market cap is impressive, in essence it is still just a marketing ploy, in a Web3 context. If we peel away the shiny acronyms of DAO, NFTs, decentralization, and cryptocurrencies, how is the SOS-OpenSea airdrop that different from a skincare brand mailing everyone who has ever left a cell number with the company a free tester of its latest products?

Although the OpenDAO campaign was short-lived, the community has shown OpenSea what they are capable of without the company's benediction. SOS may well be forgotten, but it's the first step toward the debate on "centralized" service platforms. As a result, LooksRare, a new NFT marketplace, was created with more creator-friendly protocol to compete with OpenSea.

The truth is, however, the Web3 is predicated on the existence of certain key infrastructure, because not every user wants to write code every time for a transaction, run their own servers, or worry about their crypto asset security themselves. Therefore, "centralized" services will inevitably rise, as seen in Moralis ("build, host, and scale killer DApps"), Alchemy ("powering blockchain developers globally"), Infura ("ETH nodes as a service"), and many more examples. When more regular people join the Metaverse, the challenge for the Web3 crypto natives will be to figure out the fine balance between "centralized" and "decentralized" infrastructure.

Creator-Verse: Future Creator Economy

Despite challenges, characterized by the creator economy on the blockchain internet, the Web3, is a powerful trend to come. DAOs discussed in this chapter, including ConstitutionDAO,

YGG, and OpenDao, are its important experiments. Collectively, they mark a substantial move towards a functioning open economy in the Metaverse, where new ownership-sharing and decentralized governance is the core value.

We are moving into a cyberspace owned by users and builders and orchestrated with tokens, through distributed and decentralized organizations – this is Web3 for creators. Today, the digital creator economy is only a fraction of the mainstream digital economy, but its core areas are growing – they have the potential to be the heart of the future digital economy – the Metaverse. As mentioned, there are approximately 50 million content creators in the space, who consist of mostly amateurs (46.7 million), and around 2 million professionals (see **Figure 5.4** in Chapter 5, "NFTs, Creator Economy, and Open Metaverse"). That means more average users are becoming creators themselves. The trend is even more obvious as we examine the three phases of creator economy below (see **Figure 10.9**).

Phase 1: Passive Internet

In the past, the bar for becoming a content creator was dishearteningly high. You had to be signed by a record label to

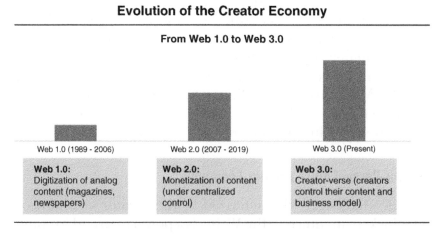

Figure 10.9   The Three Waves of "Creator Economy"

become a signer or by a publisher to become an author. If you want to be in showbiz, there are only so many television channels to choose from. And movies? A small number of studios controlled all business lines – from idea to script talent to production to theaters.

The internet in the early years (Web1) had little impact on the industry structure of creative businesses. The Web1 internet allows you to send emails, browse websites, and search information from Yahoo. It also accelerated the digitalization of contents (e.g., from paper newspapers), but it cannot help an average internet user to become an online content creator. These bottlenecks gradually dissolved with the rise of Web2 – the mobile internet. (Reference to Winston Ma's 2016 book *China's Mobile Economy*.)

Phase 2: Mobile Internet

Web2 is controlled by a handful of gargantuan companies – Big Tech. It featured the growth of social media platforms like Facebook (Meta), as well as giants like Google, Amazon, Microsoft, and Apple, who increasingly incorporate social and media contents into their software and hardware businesses. They have actively and deliberately built "moats" to trap that value and the user for as long as possible in order to extract as much "lifetime value" as possible for the benefit of such platforms (and their corporate shareholders). There is no shortage of platforms available to creators these days, but these existing platforms run contrary to the trend toward creator ownership; instead, they (not the creators) take the lion's share of the revenue from the creative products.

Phase 3: Blockchain Internet

In the Web3 world of crypto, DeFi, NFTs, and more blockchain-based technologies are converging for a paradigm oriented around the users and their sovereignty: their identity, data, creation, and wealth. While there are still platforms that help with the creation, discovery, or curation process, the user is in full control of the output and can freely transfer value between platforms to resell, borrow, and lend against in a completely permissionless way. In fact, the users would own the sites and apps in the Metaverse.

This means the creators and users will benefit the most from their creation and participation. Creating contents will result in property right. Participate in an online community enough, and you will get a piece of it in the form of a digital token. Users, rather than a large corporation, will define and govern the community and related services. Some clues may already lay in the world of gaming.

As mentioned, gaming is among the metaverse's most relevant applications and has helped pioneer revenue-generating digital marketplaces within gaming. Yes, there are metaverse-adjacent marketplaces and platforms like Fortnite and Roblox, who are dominant and centralized, and gaming developers still monetize their creation from the top-down. But the promising trend is to activate creators and monetize it from the bottom-up, as seen in P2E gaming like Axie Infinity and related DAO like YGG (see **Figure 10.10**).

Perhaps the most frequent criticism of blockchain gaming is that P2E gaming doesn't enable a user to do things that the user can't already do in games in some way. But this criticism fails to recognize that NFTs can be used to bypass traditional distribution mechanisms and enable peer-to-peer transactions, cutting out Big Tech platforms. Furthermore, there is increased financialization and better financial alignment with DeFi. Developers can share a portion of their revenue with the game's supporters by issuing tokens, and players can use their virtual items like traditional assets as collateral for usage in DeFi. (For example, a player could secure financing for their next battle pass by lending out their sword in NFT.)

Why is this important for average users? Because in the future "Creator Economy," everyone is a creator – just like gamers are also creators in **Figure 10.10**. Even before Facebook

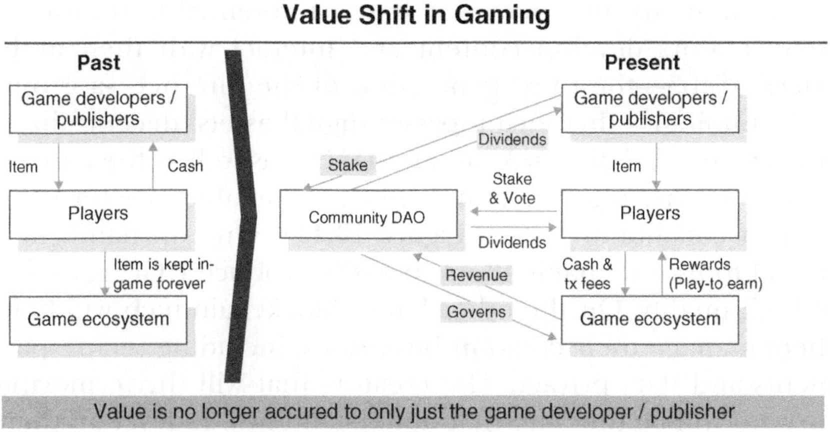

Value Shift in Gaming

Figure 10.10   Game Value Shift – Top-down vs. Bottom-up
Source: Nansen

changed its name to Meta, a large and engaged community of people who both create and consume were already on platforms like YouTube, TikTok, and Twitch. Going forward, Web3 platforms will drive exponential growth of this community, who sits at the center of the ever-growing creator economy. Now, this movement is coming for every possible thing that's "creative" from every medium.

Again, taking the gaming section as an example of this "creator expansion," the gaming players can always burn through new gaming content faster than the game makers can create it. As a result, in a high-risk, high-reward industry like games, creativity scales poorly. Talent is one of the industry's most critical bottlenecks, and the cost of making games continues to rise. Therefore, the rise of UGC (user-generated content) is not merely about user engagement, but largely a matter of necessity for gaming publishers. By opening the creative process to game players, publishers effectively outsource innovation to a huge crowd of passionate players. This enormously scales the design process, and the players (also the creators at the same time) are entitled to be compensated as a partial owner of the business.

In summary, the metaverse has vast potential to transform how creators develop content and interact with their audiences. Web3—the third generation of the internet—a group of technologies that encompasses digital assets, decentralized finance, blockchains, tokens, and DAOs (as well as the convergence of more digital technologies), will enable new forms of human collaboration (see **Figure 10.11**). One on hand, new digital tools will enable them to easily produce UGC contents at high quality. On the other hand, blockchain tech will help them manage their creation businesses, including secure payments and data privacy. The creators that will thrive moving forward will be the ones that understand how to level up their skills in their metaverse.

In this way the Metaverse provides a new digital-first economy of decentralized ledgers that is global, transparent, and

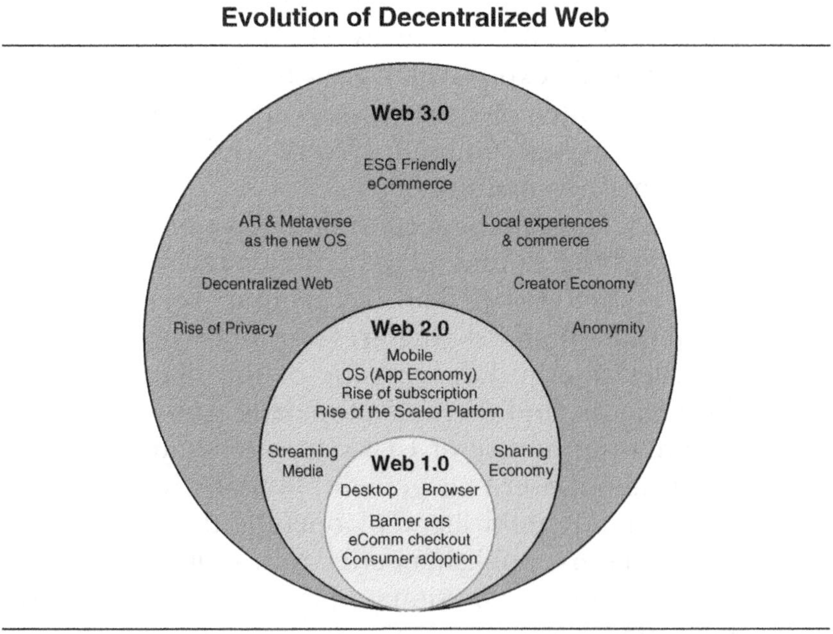

Evolution of Decentralized Web

Web 3.0

ESG Friendly
eCommerce

AR & Metaverse
as the new OS

Local experiences
& commerce

Decentralized Web

Creator Economy

Rise of Privacy

Web 2.0

Anonymity

Mobile
OS (App Economy)
Rise of subscription
Rise of the Scaled Platform

Streaming
Media

Web 1.0

Sharing
Economy

Desktop   Browser

Banner ads
eComm checkout
Consumer adoption

Figure 10.11   Evolution of the Decentralized Web, Web1.0 to Web3

Source: Goldman Sachs, "Americas Technology Framing the Future of Web 3.0" report, 2021

crypto-native. Yes, the Metaverse is facing a three-way competition with Big Tech (who are starting to fold Web3 ideas into their centralized platforms), as well as sovereign states that provide permissioned blockchain infrastructure that's cheap and convenient. And the cryptocurrency world is also competing with Big Tech tokens and the CBDC of nations. But the three-way coexistence will no doubt drive unprecedented innovation in the coming decade to build a better internet – the blockchain internet.

Glossary (alphabetical)

A

Airdrop (Crypto, NFT)   An airdrop is a distribution of cryptocurrency, tokens, or NFTs that are sent to a Web3 wallet address for free as a promotion, or as added value for participating in an experience or purchasing a digital asset. Airdrops are generally used to add additional value or to draw attention to a brand or experience.

Anti-money laundering (AML)   Activities financial institutions perform to achieve compliance with legal requirements to actively monitor for and report suspicious activities.

Atomic swap   A situation in which two parties fully exchange assets without having to trust a centralized exchange or third party. In an "atomic" transaction in digital currency, if one leg of a transaction that involves payment for an asset fails, the whole transaction fails.

Augmented reality (AR)   a technology that superimposes a computer-generated image on a user's view of the real world, thus providing a composite view.

Adverse selection   Market situation where buyers and sellers have different information. Users with more information participate selectively in trades when they deem it profitable, decreasing the quality of market for everyone.

AMM (automated market maker)   An automated market maker uses a pair of assets in a pool that are deposited by a liquidity provider. A trader can then trade one asset within the pool for the other, paying a fee. The price will fluctuate with demand along a liquidity curve. Popular examples of automated market makers are Uniswap, Sushiswap, and Pancakeswap.

API (application programming interface)   An API provides an end point for developers to connect to so they can gain access to data and execute functions programmatically. Exchanges will provide API access and API keys for their users so they can trade programmatically using trading bots and scripts.

APR / APY (annual percentage rate / yield)   APR represents the annual percentage rate charged or earned for borrowing or lending money. However, this doesn't take into account the effect of compounding. If interest is paid out monthly, the lender will earn interest on their interest. This compounding effect is taken into account using the APY calculation but not with APR.

Address (wallet address)   A blockchain wallet address such as Bitcoin and Ethereum wallet address is a synonym for a public key. It's the address that you share with someone so they can send funds to your wallet.

Audit   A security audit is performed by an external organization on a project's smart contract code. It provides some reassurance but by no means guarantees the safety of funds within a smart contract. Not all auditors are created equally; an audit by a leading firm such as Certik, KnowSec, and Slowmist carries more weight.

B

Beeple, a.k.a Mike Winkelmann   American graphic designer who does a variety of digital artwork including short films, Creative Commons VJ loops, everydays, and VR / AR work. One of the originators of the current "everyday" movement in 3D graphics, he has been creating a picture every day from start to finish and posting it online for over 10 years without missing a single day.

Bitcoin   Bitcoin is a decentralized digital currency, without a central bank or single administrator, that can be sent from user to user on the peer-to-peer Bitcoin network without the need for intermediaries. The currency is a bitcoin.

Block   Blocks are data structures within the blockchain database, where transaction data in a cryptocurrency blockchain are permanently recorded. A block records some or all of the most recent transactions not yet validated by the network. Once the data are validated, the block is closed.

Blockchain   A form of distributed ledger technology (DLT) in which transactions are conducted in a peer-to-peer fashion and then broadcast to the entire set of system participants, all or some of whom work to validate them in batches known as blocks. Such validation is executed using the system's consensus protocol (such as proof-of-work or proof-of-stake). Validated blocks are then cryptographically linked to a primary sequence of blocks, referred to as a blockchain.

Block confirmation   Exchanges and payment protocols often implement a minimum number of block confirmations to deposit funds. Each time a miner finds a hash and the block is finalized, it counts as a block confirmation. So if a transaction requires three block confirmations, this will be the block that contains your transaction plus two more on top to be completed.

Block height   The number of blocks within a blockchain. This is often used as a de facto timing mechanism within smart contracts as developers can estimate the block height at a particular time in the future based on the average block times.

Block reward   Block reward includes the mining fees and any transaction fees paid to miners when they find a hash that meets the difficulty rating. Each block will carry a reward for helping secure the blockchain, which is how many cryptocurrencies distribute the supply of the token.

Block producer   A block producer (BP) is a person or group whose hardware is chosen to verify a block's transactions and begin the next block on most proof-of-stake (PoS) blockchains.

Bonding curve A mathematical formula or curve used to define a relationship between price and supply of an asset. Bonding curve contracts are used by some projects to increase the price of a token being sold as the supply increases.

Blockchain-based Service Network (BSN) China's state-backed multi-blockchain hosting network initiated by the State Information Center along with Red Date, China Mobile, and China UnionPay. It is a common infrastructure for the deployment and operation of blockchain applications globally.

C

Central Banking Digital Currency (CBDC) A digital form of central bank money that may be accessible to the public (general-purpose or retail CBDC), or to a select set of licensed participants such as financial organizations (wholesale CBDC). CBDC is denominated in the national unit of account. It is issued by and is a direct liability of the central bank.

Centralized exchange A business service that acts as an intermediary in an exchange transaction to enable the conversion to and from certain assets or currencies.

Confidentiality Relates to the ability to keep certain information private from nonpermitted parties. Confidentiality in some legal systems is protected by a duty on the recipient not to divulge to third parties without the discloser's consent. It is also sometimes protected by agreement between the discloser and recipient.

Consensus mechanism Consensus mechanisms (also known as consensus protocols or consensus algorithms) allow distributed systems (networks of computers) to work together and stay secure.

Consortium Chain (aka Federated Blockchain) Blockchain technology where instead of only a single organization, multiple organizations govern the platform. It's not a public platform but a permissioned platform.

Crypto-assets An asset that heavily involves the use of cryptography and that operates on a distributed ledger. Cryptocurrencies such as bitcoin and ether are examples. However, "crypto-assets" is a broad term that can also include other assets that exist and can exchange hands on a distributed ledger.

Crypto token A type of cryptocurrency that represents an asset or specific use and resides on their own blockchain. Tokens can be used for investment purposes, to store value, or to make purchases.

Cold wallet / Cold storage A cold wallet is a way of storing funds by keeping the private key offline. An example would be when the Winklevoss twins, who were early adopters for Bitcoin, purchased a laptop, set up a private key/public key pair, divided the private key into three parts, each part was duplicated, and then put each of the six parts in a different bank security box across the country. This is an extreme example of keeping keys secure. The main idea is to make it (almost) impossible for hackers to gain access to your keys if they are not connected in any way to the internet.

Collateral When taking out a futures position or borrowing funds on a lending platform, something of value (collateral) is used to secure the loan.

Composability In DeFi terms composability is the potential for smart contracts that form the DeFi protocols to interact with each other. A contract might connect to a lending platform to take out a flash loan and then use those funds to interact with an automated market maker to swap tokens for example.

Compound interest When a user invests a sum of money an annual rate of interest will often be quoted. However interest is usually paid out more regularly, sometimes as often as every block. This means that the interest we get paid today will start earning interest itself tomorrow.

Contract address   A smart contract address is like the post code of a smart contract on a decentralized network. It maps to the memory address of the executable code on the virtual machine. When we want to interact with a contract, we often need the contract address. A common example of this is a token address that describes where to find that token contract.

D

Decentralized autonomous organization (DAO)   A DAO is a type of formalized community in crypto in which members have their incentives aligned through mutually agreed upon governance mechanisms and, often, through the use of a specific cryptocurrency or token.

Decentralized App (DApp)   Digital applications or programs that exist and run on a blockchain or peer-to-peer (P2P) network of computers instead of a single computer. DApps (also called "dapps") are outside the purview and control of a single authority.

Decentralized finance (DeFi)   An umbrella term for Ethereum and blockchain applications geared toward disrupting financial intermediaries.

Delivery versus payment (DvP)   A settlement mechanism that ensures that the final transfer of an asset, namely an investment security, occurs only if the final transfer of payment for the asset takes place. DvP transfers can occur within a jurisdiction or across borders.

Digital currency   Typically used to refer to currency that exists in electronic form and that may or may not be available in physical form. Digital currencies often have some characteristics of a currency, namely serving as a store of value, unit of account, or medium of exchange, although the term may also be used more liberally. They may also have characteristics of a commodity or other asset.

Digital identity   A set of digital credentials used to represent and prove the identity of a real-world individual, organization, or electronic device on electronic or online systems, and their right to access, for example, certain information and services. Today, these typically take the form of digital certificates created using public-key cryptography to bind together a public-key with identity details and other details, such as a private key and the owner's digital ID.

Decentralized Identifier (DID)   A new type of identifier that is globally unique, resolvable with high availability, and cryptographically verifiable. DIDs are typically associated with cryptographic material, such as public keys and service endpoints, for establishing secure communication channels.

Digital Silk Road (DSR)   The Digital Silk Road is the technology dimension of China's Belt and Road Initiative. It is advancing in several areas: wireless networks, surveillance cameras, subsea cables, and satellites.

Digital token   A unit on a digital and typically decentralized ledger that is used to represent value, such as an asset or a basket of assets, including real-world assets such as commodities, stock, or real-estate property. The token can be used to facilitate transactions and transfers of title to such underlying value or asset.

Digital wallet   A digital device, software-based system, or online application for storing payment information such as passwords and private keys, which when used in conjunction with a payment system can enable online payments. When they involve cryptocurrency, digital wallets are also used as a mechanism to store private key information for users to access their cryptocurrencies.

Distributed ledger technology (DLT)   An overarching term that includes blockchain technologies and refers to the protocols and supporting infrastructure that allow computers in different locations to propose and validate transactions on a ledger and update ledger records in a synchronized way across a network. Many DLTs are designed to function without a centralized trusted authority, relying instead on distributed consensus-based validation procedures combined with cryptographic signatures.

DEX (Decentralized Exchange)   Decentralized exchanges include orderbook based exchanges like IDEX and automated market makers like Uniswap. An orderbook exchange will list bid and ask prices, and users will be able to place orders into the book, which are filled by a matching engine. An automated market maker uses a liquidity pool of two assets, which can then be traded against the pool along a price curve.

Decentralization   The entire blockchain sector is built around the concept of decentralization. This means that a network has no central point of failure and is instead built around equal peers. Decentralization is not a binary concept; networks can become more or less decentralized over time.

Deflationary token   A deflationary token is an asset where the circulating supply reduces over time. It becomes more rare often through a burning process where tokens are sent to an address that no one has access to.

Degen   Short for degenerate, which in DeFi terms can be used both as an insult and a compliment at the same time. It is usually assigned to a trader, yield farmer, or NFT collector who takes on high risk strategies. Someone who trades meme coins with their life savings on leverage would be considered a degen.

Delegated proof of stake (DPoS)   DPoS is a consensus algorithm where stakers can allocate their voting capacity to third-party nodes on the network. It removes the need for stakers to run nodes themselves, as they can simply vote through a node operator providing trust in that party acting in their interest to secure the blockchain.

Derivative   A financial instrument that is used to gain exposure to an underlying asset. In crypto, the most popular example is that of perpetual futures contracts. Quarterly futures and options contracts are other forms of derivatives. In crypto markets derivatives are traded at greater volumes than the underlying spot markets. This means there is more buying and selling of Bitcoin futures than there is of actual Bitcoin.

Digital Signature   Transactions need to be signed before they are sent to the nodes that form the blockchain network. This is achieved via a private key or public key pair and is usually done in the background via a digital wallet such as MetaMask. The transaction data will be hashed and then signed using the private key and elliptic curve cryptography. This will then be sent to nodes along with the public key to prove the sender approves the transaction. The nodes have a cryptographic function to check if the signature matches the public key for the account.

Double spend   Bitcoin's initial breakthrough was to solve the double spend problem, which ensures a user on a decentralized network can't send their coins to different addresses on different peers. The consensus mechanism ensures that only one block will move forward, and that can only include a single spend of the tokens.

E

Electronic money (E-money)   Stored value held in digital accounts or physical devices (e.g., a chip card or a hard drive in a personal computer) that is used as a means of payment and a store of value. E-money systems vary across different jurisdictions, but they are often fully backed by fiat currency, denominated in the same currency as central bank or commercial bank money and exchangeable at par value for such money or redeemable in cash.

Ethereum (ETH)   Ethereum is a blockchain platform with its own cryptocurrency, called ether (ETH) or Ethereum, and its own programming language, called Solidity.

ERC20 token   The majority of crypto tokens use an ERC20 token contract. Anything that is traded on Uniswap or Sushiswap is ERC20 or a variation built on top of it. The ERC20 token contains functions to create, transfer, approve spend, and check balances.

ERC777 token   Like ERC20, ERC777 is a standard for fungible tokens, and is focused around allowing more complex interactions when trading tokens. More generally, it brings tokens and ether closer together by providing the equivalent of a msg.value field, but for tokens.

ERC721 token   The ERC721 token is the industry standard token used for NFT's. It contains many of the standard ERC20 token functions alongside additional functions to declare and modify ownership and store metadata. Metadata contains the data, which the NFT represents; it is often a hash of the data rather than the data itself.

ERC1155   A novel token standard that aims to take the best from previous standards to create a fungibility-agnostic and gas-efficient token contract. ERC1155 draws ideas from all of ERC20, ERC721, and ERC777.

Etherscan   A block explorer provides a user interface for anyone to search for transactions, user accounts and blocks on a blockchain network. Etherscan is Ethereum's block explorer and is a pillar of the industry. When users send a transaction, they'll often be quoted a confirmation tx address, which can be copied-and-pasted into etherscan to see the details of that transaction.

Ethereum Virtual Machine (EVM)   A virtual machine is like a version of windows running in a window on your laptop. Think of it as an operating system running as an application on top of the main operating system. Ethereum's virtual machine is designed to run across a network of nodes that agree on the persistent state of data on the network. It's not just Ethereum that uses EVM; it's also used by alternate chains like Binance smart chain, Polygon, and Avalanche.

F

Fiat currency   A form of currency established by government decree and generally issued by a monetary authority such as a central bank. Fiat currencies can be distinguished from other historic forms of government-issued money by typically not being backed by a commodity such as gold or silver. Fiat currency can take the form of physically issued bank notes and cash or it can be represented electronically, such as with bank credit, central bank reserves, or central bank digital currency (CBDC).

FOMO   Fear of missing out. It's the feeling you get when you work the industry only to find out your Uber driver has outperformed your portfolio because he invested in a meme token that went viral on TikTok. FOMO can lead us to invest at the worst possible time when markets are toppy and due for a correction.

Fair launch   The concept of a fair launch token was popularized by Yearn Finance when it released its governance token without any team allocation or VC interest. It simply gave it away to the people that were using the protocol. This created a strong community that benefits from the project to this day.

Financial primitive   Simple financial products such as loans and insurance can be classed as financial primitives. They are the fundamental financial services that a protocol may provide. In a DeFi sense, financial primitives are often used to describe the complete ecosystem around which a token economy is built.

Flash loan   The concept of a flash loan is quite abstract in that it lets a user borrow millions of dollars with no collateral but only for a few seconds. A flash loan must be paid back in the same block that it is borrowed or the transaction will fail.

Fork (hard fork / soft fork)   The blockchain sector prides itself in being transparent, which includes the vast majority of code being open source. This means that it can be forked, copying existing code to our own project and then modifying it from there. When major changes are pushed out, a subset of the nodes may not accept them continuing with alternate or preexisting code. This division of nodes is known as a hard fork. An example of this took place on Bitcoin where Bitcoin Cash split off due to a debate over block sizes.

G

GameFi   In its most common usage, GameFi refers to decentralized applications ("dapps") with economic incentives. Those generally involve tokens granted as rewards for performing game-related tasks such as winning battles, mining precious resources, or growing digital crops. It's an approach also known as play-to-earn.

Gameplay   Features of a video game specifically contributing to the gaming experience it offers to its users.

Gas fees   Gas refers to the fee, or pricing value, required to successfully conduct a transaction or execute a contract on the Ethereum blockchain platform.

Global Payment Rail   Global real-time instantaneous settlement and clearing for payment enabled by cryptocurrencies.

Genesis block   The first block on a blockchain is known as the genesis block. Bitcoin's genesis block famously included an encoded message saying "The Times 03/Jan/2009 Chancellor on brink of second bailout for banks."

Gwei   Ethereum's native token ether (ETH) can be broken down into one billionth denominations known as gwei. These are used more in development than on user interfaces. 1 ether = 1,000,000,000 gwei.

H

Hash   A hash is a function that meets the encrypted demands needed to solve for a blockchain computation. Hashes are of a fixed length since it makes it nearly impossible to guess the length of the hash if someone was trying to crack the blockchain. The same data will always produce the same hashed value.

HODL   In December 2013 BitcoinTalk forum user Game-Kyuubi posted a somewhat intoxicated message declaring "I AM HODLING." Having misspelled *holding*, the post led to the term sticking and is now widely used within the community. It simply means to hold onto an asset through the ups and downs. Often quoted as hold on for dear life.

Halving   Halving events occur when a token's distribution of new supply to miners is cut in half. This occurs approximately once every four years on the Bitcoin network with the next halving due in 2024.

Hardware wallet   A small USB type device that stores private keys and the funds associated with them in a secure manner. It can often be disconnected completely from the internet, making it more difficult for hackers to gain access.

I

The Internet of Things (IoT)   Network of physical objects – "things" – that are embedded with sensors, software, and other technologies for the purpose of connecting and exchanging data with other devices and systems over the internet.

Interoperability   Interoperability refers to the basic ability of different computerized products or systems to readily connect and exchange information with one another, in either implementation or access, without restriction.

Immutability   Blockchains are immutable because no one is able to change the existing data. Blocks are interlinked and stacked on top of each other with each new block containing a hash of the underlying block. Changing a block from three days ago would mean every block since would need to be recalculated and rewritten.

Impermanent loss When a liquidity provider deposits funds to an automated market maker they receive fees in exchange for accepting the risk of impermanent loss. If one asset goes up in price and the other goes down, the pool will fill up with the lower value asset. The liquidity provider is always on the bad end of price action. If the price returns back to the base level, such as often is the case with stablecoins, no impermanent loss will be suffered, however, if the price move is permanent so is the loss.

K

Know your customer (KYC) Processes and protocols, usually prescribed by law, that apply to certain accountable institutions, such as banks, obliging them to verify and keep records of the identities of their customers in line with strict global or national anti-money laundering, anti-terrorism, and other laws and regulations.

L

Layer 2 L2's are sub-chains that form consensus based on smart contracts, which live on the layer 1 main chain. Optimistic rollups are an example of layer 2 scaling solutions, which promise faster, cheaper transactions with the benefit of layer 1 security.

Leverage When a trader makes a trade with leverage they are effectively borrowing money to place that trade. If the trade goes against them, they risk being liquidated if the loss comes close to exceeding their collateral position. For example, a user can deposit $100 to an exchange, purchase a Bitcoin futures position worth $2,000 with 20x leverage, but if the price of Bitcoin drops close to 5 percent, they risk getting liquidated and losing their deposit.

Liquidation   When using leverage it's important for the protocol or exchange to prevent losses exceeding the collateral posted. For this reason, a liquidation engine will sell positions to recap funds automatically if a margin requirement is not met. Liquidation engines work differently across the industry but many market sell assets, which can cause liquidation cascades and highly volatile price action.

Liquidity mining   Protocols often require funds to operate. For example, a lending and borrowing platform needs a float and lenders before they can start lending. DeFi protocols will often bootstrap initial funding through liquidity mining. This is the incentivization to get users to deposit funds to the platform. This may take the form of distributing governance tokens to early adopters or providing high APY returns for staking LP tokens for the ETH/Native pair providing a liquid market for the governance token.

Liquidity pool   A liquidity pool usually contains a pair of assets that can be swapped. For example a Uniswap liquidity pool might have ETH as the base asset and an ERC20 Token as the traded asset. Price is calculated along a curve dependent on the quantity of assets in the pool. If someone starts buying the ERC20 token with ETH, it pushes the price up as more ETH is added and the ERC20 tokens are removed from the pool.

Liquidity provider   A liquidity provider will usually provide a pair of assets such as ETH and ERC20 tokens in equal weighting to a liquidity pool. They will earn fees whenever someone trades in that liquidity pool. When providing liquidity they will receive LP tokens in return (see below).

LP tokens (liquidity provider tokens)   LP tokens act like a receipt for the funds deposited, and they will automatically be sent to the same address that deposited the funds. LP tokens can be transferred and can often be staked on DeFi platforms in return for staking rewards.

M

MakerDAO   MakerDAO enables the generation of Dai, the world's first unbiased currency and leading decentralized stablecoin.

Metadata   Metadata is defined as the data providing information about one or more aspects of the data; it is used to summarize basic information about data, which can make tracking and working with specific data easier. There are three main types of metadata: descriptive, administrative, and structural. A simple example of metadata for a document might include a collection of information like the author, file size, the date the document was created, and keywords to describe the document. Metadata for a music file might include the artist's name, the album, and the year it was released.

Metaverse   It is a combination of multiple elements of technology, including virtual reality, augmented reality, and video where users "live" within a digital universe. Supporters of the metaverse envision its users working, playing, and staying connected with friends through everything from concerts and conferences to virtual trips around the world.

Mining   The process that bitcoin and several other cryptocurrencies use to generate new coins and verify new transactions. It involves vast, decentralized networks of computers around the world that verify and secure blockchains – the virtual ledgers that document cryptocurrency transactions.

Mainnet   Mainnet is the term for the real blockchain network, and is used in contrast with testnet networks. Unlike the other networks, which are used for testing purposes, mainnet coins have monetary value.

Margin   A trade made on margin is executed using borrowed money. A percentage of the total trade value is kept on account as collateral to cover potential losses. If losses exceed collateral a liquidation event will occur and the trader will lose the collateral posted.

Market cap   The market cap or capitalization of a cryptocurrency is calculated by multiplying the circulating supply by the token price. This is usually a debatable issue with leading websites not including vested tokens and Treasury wallets in the circulating supply.

Market maker   A market maker will provide liquidity to an order book on a traditional exchange. They will often place both bid and ask to buy and sell the same asset at a varying spread away from the current market price.

Maximalist   Maximalism is a mindset in which someone feels that a single coin or token holds value above everything else. Bitcoin maximalism arose toward the end of 2017 with maxi's declaring everything else in the sector worthless. More recently we've seen more Ethereum maximalism where proponents believe that alternate chains are meaningless.

Merkle tree   Merkle trees are a data structure where hashes are used for verification. A root hash can be used to verify underlying blocks of data provided across an untrusted peer-to-peer network. It's possible to verify each block of data contains the commitment from the root hash.

MEV   Miner extractable value (MEV) is a measure of the profit a miner can make through their ability to arbitrarily include, exclude, or reorder transactions within the blocks they produce.

Money Lego   Money Lego are tech stacks that allow different applications to fit (or be shoved) into other projects. For example, you can deposit ether (ETH) into MakerDAO, receive the stablecoin dai (DAI), and then lend it on Compound to a trader in order to earn the network's governance token COMP.

Multi signature wallet (MultiSig) A multisignature wallet or multisig is a digital wallet that requires multiple signatures to transfer funds. For example, a Gnosis multisig wallet might be set up by a team who want to secure their Treasury funds. There might be five team members who are signatories on the account, and it may be set to require at least three signatures for a transaction. Each user will be given a private key/public key pair via a digital wallet like MetaMask. They can then propose and sign any transactions to transfer funds, which won't go through until three team members have signed off on the transaction. Multisig wallets are used to mitigate the risk of theft, lost keys, and hacked funds.

N

Node A node is a computer connected to other computers, which follows rules and shares information. A full node is a computer in bitcoin's peer-to-peer network that hosts and synchronizes a copy of the entire bitcoin blockchain. Nodes are essential for keeping a cryptocurrency network running.

Nonfungible token (NFT) Unique and noninterchangeable unit of data stored on a digital ledger. NFTs can be associated with reproducible items such as photos, videos, 3D models, audio, and other types of digital files as unique items. NFTs use blockchain technology to provide a public proof of ownership.

O

Oracle Smart contracts cannot connect to external data sources such as APIs. For this reason, to get information into a contract, it must be provided by a service such as an oracle. Oracles can provide any type of data but in DeFi it is usually price data from centralized exchanges. This is useful for developers to prevent the risk of price manipulation on-chain.

P

P2E (play to earn) A new business model for gaming companies, where users are rewarded for their participation either with an in-game currency generated by the platform or with unique items within the game. The ownership of both these assets is recorded on-chain through token standards, either fungible or not.

Payment versus payment (PvP) A settlement mechanism that ensures that the final transfer of a payment in one currency occurs only if the final transfer of a payment in another currency or currencies takes place. PvP transfers can occur within a jurisdiction or across borders.

Peer-to-peer (P2P) Refers to interactions between peers in a system, such as transactions or information exchange, which occur without the need of an intermediary. In the blockchain industry, this has come to refer to systems that enable transfers of value without an intermediary bank, utilizing, for example, distributed ledger technology.

Professional generated content (PGC) Content generated by the brand itself in order to let people know its brand and much more that they have to offer through images, videos, blog posts.

Proof-of-stake (PoS) Proof-of-stake is a cryptocurrency consensus mechanism for processing transactions and creating new blocks in a blockchain.

Proof-of-work (PoW) A piece of data that requires a significant amount of computation to generate but requires a minimal amount of computation to be verified as being correct. Bitcoin uses proof of work to generate new blocks.

Protocol Protocols are basic sets of rules that allow data to be shared between computers. For cryptocurrencies, they establish the structure of the blockchain – the distributed database that allows digital money to be securely exchanged on the internet.

Privacy-enhancing technology (PET) Technologies or systems that incorporate technical processes, methods, or knowledge to achieve specific privacy or data protection functionality, or that implement specific requirements of data protection laws and reduce the risks associated with processing personally identifiable information, such as the risk of data breaches.

Public key infrastructure (PKI) The policies, procedures, software, and hardware required to create, manage, distribute, use, store, and revoke public and private key pairs and digital certificates that are used for encryption and other purposes. The public key can be openly shared to relevant parties without compromising security, while the private key must be kept confidential. Private keys are typically required to decrypt confidential information and messages. They can also be used to create a digital signature on a message or document. A digital signature is a mathematical scheme that demonstrates to the recipients that the message or document in question originated with the private key's owner and that there has not been forgery or tampering.

Professional user-generated content (PUGC) PUGC focuses on the influencer, enabling them to create intimacy with their audiences. A marketing team maintains the influencer's original video content while managing the business collaborations and product placements. It hinges on the fact that the team gives the illusion that the content is done completely by the individual influencer alone.

Private key A set of ones and zeros often represented in hexadecimal alphanumeric format. It acts as the primary data input for account creation in cryptocurrency because the public key is derived from the private key. Private keys, as the name suggests, should be kept private as anyone who has access can sign transactions and take any funds in the account.

Q

Quantum computing Is a type of computation that harnesses the collective properties of quantum states, such as superposition, interference, and entanglement, to perform calculations. The devices that perform quantum computations are known as quantum computers.

Quantum supremacy In quantum computing, quantum supremacy, or quantum advantage is the goal of demonstrating that a programmable quantum device can solve a problem that no classical computer can solve in any feasible amount of time.

Quantum resistant blockchain Quantum resistant blockchain utilizes quantum mechanics and cryptography to enable two parties to exchange secure data and detect and defend against third parties attempting to eavesdrop on the exchange. The technology is seen as a viable defense against potential blockchain hacks that could be conducted by quantum computers in the future

R

Real-time gross settlement (RTGS) In the context of interbank settlement, RTGS refers to systems for the continuous and real-time transmission of funds or securities individually on an order-by-order basis, without netting.

Retail CBDC A form of central bank digital currency (CBDC) that is accessible to the general public. Retail CBDCs may take a two-tiered structure, where citizens would hold CBDC balances with commercial banks or other customer-facing financial entities, such as private payment service providers, rather than directly with the central bank. A retail CBDC could be used both domestically and cross-border (i.e., accessible and usable by foreign entities). Retail CBDCs are sometimes also referred to as general purpose or universally available CBDCs.

Rollups A form of layer two-scaling solution. Transactions are rolled up in an amalgamation process and stored in an inbox within a layer 1 smart contract. The transactions are processed via external nodes on layer 2, taking a lot of the execution and computational work away, then the state is updated and sent back to layer 1. A dispute mechanism is used to prevent misuse between validator nodes on layer 2.

S

Self-sovereign identity (SSI) An assembly of principles stating that the individual should own and control their identity without the intervening third party and centralized authorities. SSI is built around three pillars: Security, controllability, and portability. Under this paradigm, identities are owned and controlled by the entity they represent.

Smart contract Self-executing contract with the terms of the agreement between buyer and seller being directly written into lines of code.

Stablecoin A digital currency that is pegged to a "stable" reserve asset like the US dollar or gold. Stablecoins are designed to reduce volatility relative to unpegged cryptocurrencies like bitcoin.

Sustainable token economy Ponzi scheme-like systems that use later arrival participants' funds to pay for early participants can only survive a short duration of time and will not be sustainable over a longer period.

Synthetic CBDC Refers to an alternative framework to central bank digital currency (CBDC), under which private payment service providers hold reserves at the central bank that fully backs the digital currency they issue to customers. The regulatory framework would intend to guarantee that these providers' liabilities will always be fully matched by funds at the central bank, creating protection for users against issuer default. Such liabilities could share some of the characteristics of a CBDC issued by the central bank, but they could not constitute CBDC, as the end-user would not hold a direct claim on the central bank. Synthetic

CBDC is neither issued by nor a direct liability of the central bank. Synthetic CBDCs have been referred to as a form of "narrow-bank" money.

Shard, sharding   A shard is a subset of data, and sharding is used by data management software to break down large data sets into more manageable packages. By the end of 2021 the Ethereum blockchain will be over 1TB in size, and transferring this data across a decentralized network potentially could become more difficult. Sharding will enable nodes to work with a subset of the entire blockchain, which will ease the computational burden of past transactions.

Slippage   The price movement caused by an order. When an asset is traded on exchange, the quoted price is often the midpoint between the leading bid and leading ask price. However when a market order is placed it can take out more than just the leading price eating into the order book and removing liquidity.

Solidity   The main coding language used to create smart contracts on the Ethereum network. It's a statically typed language designed around the Javascript syntax making it familiar for web developers.

Staking   DeFi protocols will often incentivize funding and liquidity providers by distributing a governance token to staked funds. A user can either use the protocol or purchase the governance token on exchange and use this to stake and earn further funds.

Sub-chain / side-chain   Ethereum is open source code, which means it can be forked and changed by anyone who understands how it works. Sub-chains like Polygon are modified copies of the Ethereum code that run a separate chain in parallel. Some protocols will deploy smart contracts across multiple side-chains.

Synthetic assets (Synths)   Synthetic assets are a derivative product that aims to track an underlying asset. A user can trade stocks, index funds, commodities, and cryptocurrencies using synthetic assets. They are backed by a liquidity pool that acts as a balancing and funding mechanism for the protocol. If all the synthetic assets go up at the same time, then the liquidity pool diminishes in value. In practice, the diversified nature of the assets works well to keep things in balance.

T

Token A "token" often refers to any cryptocurrency besides bitcoin and ether. It also describes crypto assets that run on top of another cryptocurrency's blockchain.

Total value locked(TVL) The value of collaterals or assets locked in a game.

Tokenomics For a token to go up in value, the demand must outweigh the supply on exchange. The economics of the token ecosystem are known as tokenomics. There are various methods to try and increase demand and reduce supply, such as staking, fee burning, and holder benefits.

Testnet A playground for developers and end users to try out things with valueless funds. For example, a developer can get free testnet ETH from a faucet and use this to deploy smart contracts. A user can use a free ETH to try out new DeFi platforms and experiment with the latest innovations without risking any funds.

TradFi Traditional finance or centralized finance. The institutions of Wall Street and centralized stock exchanges would be considered TradFi.

U

User-generated content (UGC) Any content – text, videos, images, reviews, etc. – created by ordinary internet users, rather than industry professionals.

V

Virtual reality (VR) VR uses cutting-edge graphics, best-in-class hardware, and artistically rendered experiences to create a computer-simulated environment where users are not just a passive participant but a co-conspirator.

Validator A user can stake their tokens on a proof of stake network and run a node to actively participate in the validation of the blockchain. Validator nodes connect to the peer-to-peer network to process transactions and blocks.

Volatility Many cryptocurrency assets are described as being highly volatile. This means that the price can swing wildly in both directions. Bitcoin often has 50%+ drawdowns, and altcoins are even more volatile. In 2018, many tokens lost 95%+ of their USD value causing disruption throughout the industry.

W

Web2.0 Web2.0 is the second stage of development of the World Wide Web, characterized especially by the change from static web pages to dynamic or user-generated content and the growth of social media.

Web3 Web3 represents the next iteration or phase of the evolution of the Web/internet and could potentially be as disruptive and represent as big a paradigm shift as Web2.0. Web3 is built on the core concepts of decentralization, openness, and greater user utility.

Wholesale CBDC A form of central bank digital currency (CBDC) that would be used among licensed banks and other financial institutions that typically hold reserve deposits with a central bank for interbank payments and securities transactions. Wholesale CBDC could be used both domestically and cross-border. Domestic wholesale CBDC is akin or equivalent to the reserve accounts commercial banks often hold with central banks today.

Whale A crypto whale is an affectionate term used to describe someone who has a very large holding in cryptocurrency. These are generally early adopters, crypto funds, and high-net-worth individuals.

Y

Yield The return on investment we can get from staking or lending. This is usually provided by a platform as an APR figure or APY figure (includes compound interest).

Yield aggregator   A yield aggregator will automate some of the yield farming process by claiming staking rewards and then restaking to compound the returns.

Yield farming   Yield farmers operate a cat and mouse game of looking for new protocols to get in on early and start earning the best rewards. Often, yield farmers will only participate in a single farm for a period of a few days or weeks before switching to the next project that wants to bootstrap liquidity and is offering incentives to do so.

Z

Zero-knowledge proofs (ZKPs)   Zero-knowledge proofs use cryptographic methods to verify data without sharing the actual data. In cryptocurrency, ZKPs can be used to validate a transaction without revealing whose wallet was used to send the funds and how much was sent. This adds the potential for a privacy aspect to an otherwise transparent blockchain system.

****

5G   5G is the fifth-generation mobile network. It is a new global wireless standard after 1G, 2G, 3G, and 4G networks. 5G enables a new kind of network that is designed to connect virtually everyone and everything together including machines, objects, and devices. 5G wireless technology is meant to deliver higher multi-Gbps peak data speeds, ultra-low latency, more reliability, massive network capacity, increased availability, and a more uniform user experience to more users. Higher performance and improved efficiency empower new user experiences and connects new industries.

Index

Printed and bound by CPI Group (UK) Ltd, Croydon, CR0 4YY

18/11/2022

03164342-0001